解码智能时代 2021

从中国国际智能产业博览会瞭望全球智能产业

黄桷树财经　编著

撰稿：张山斯　胡　浩　何永红　骆小燕　姚　伟
　　　李全龙　吴荣飞　甘　勤　孙小梅

翻译：王　蓉　胡文江　胡　宇　石青枚
　　　孟　浩　张丽荣　赵　蔺　程浩源

重庆大学出版社

图书在版编目（CIP）数据

解码智能时代.2021.从中国国际智能产业博览会瞭望全球智能产业 = Decrypting the Intelligent Era 2021: Overlook the Global Intelligent Industry from Smart China Expo：汉、英 / 黄桷树财经编著；王蓉，胡文江译. -- 重庆：重庆大学出版社，2021.8（2021.8重印）
ISBN 978-7-5689-2864-9

Ⅰ.①解… Ⅱ.①黄…②王…③胡… Ⅲ.①人工智能—汉、英 Ⅳ.①TP18

中国版本图书馆CIP数据核字(2021)第140800号

解码智能时代2021：从中国国际智能产业博览会瞭望全球智能产业
JIEMA ZHINENG SHIDAI 2021: CONG ZHONGGUO GUOJI ZHINENG CHANYE BOLANHUI LIAOWANG QUANQIU ZHINENG CHANYE

黄桷树财经　编著
王　蓉　胡文江　译
策划编辑：雷少波　许　璐　杨　琪
责任编辑：许　璐　杨　琪　　版式设计：许　璐
责任校对：夏　宇　　　　　　责任印制：张　策

＊

重庆大学出版社出版发行
出版人：饶帮华
社址：重庆市沙坪坝区大学城西路21号
邮编：401331
电话：（023）88617190　88617185（中小学）
传真：（023）88617186　88617166
网址：http://www.cqup.com.cn
邮箱：fxk@cqup.com.cn（营销中心）
全国新华书店经销
重庆俊蒲印务有限公司印刷

＊

开本：720mm×960mm　1/16　印张：19　字数：337千
2021年8月第1版　　2021年8月第2次印刷
ISBN 978-7-5689-2864-9　　定价：88.00元

本书如有印刷、装订等质量问题，本社负责调换
版权所有，请勿擅自翻印和用本书
制作各类出版物及配套用书，违者必究

前言

盛会 趋势 实践 成果 重庆

　　本书不是对中国国际智能产业博览会（以下简称"智博会"）的简单记录。

　　而是以智博会为窗口，带领读者感受以智能化为代表的数字经济如何改变世界，数字如何产业化，产业如何数字化，现在正发生着什么，将来会发生什么，是一次思维的体操，是一次思想的盛宴。

　　本书分"盛会""趋势""实践""成果""重庆"五章。

　　"盛会"：对2020线上智博会的记录与介绍。突出对创新思想的记录，展现全球精英在智能化方向的思想演进。

　　"趋势"：展望了智能产业的发展趋势。未来已来，大势已明。

　　"实践"："趋势"空中展开、"实践"具体落地，聚焦制造业、农业、城市建设、公共卫生等各种场景应用。

　　"成果"：回到读者的身边，检视人工智能已经实现的成果，智能材料、工业互联、自动驾驶、智慧文旅、智慧防疫、智慧金融、智慧政务等，不再是想象，而是可感知的现实。

"重庆"：回到重庆，站在一个全新时代重新审视，从想象、战略、实践与未来等多个视角，探究这个"智造重镇"、这座"智慧名城"，从何而来，向何处去。

本书最重要的价值在于引领读者见证历史。

我们正处于加速变化的浪尖上，这种高速发展超过了人类历史的任何时刻。但我们当中的大部分，或许还没有特别直观地感受到。就像一条身处大河里的鱼，不一定能体会到大河奔流的激荡。而这本书，就是要助推水里的鱼儿们，跃出水面，从更高更广的角度看到这个震撼的时代。

我们每一个读者，都可以见证这个大时代。

从南方古猿进化到现在，人类用了几百万年时间，而且以不同文明阶段来衡量，演化速度一直在不断加快。

狩猎文明的历程以万年为计量单位，农耕文明持续了数千年时间，工业文明只有短短几百年，而计算机及人工智能文明，发端于几十年前，眼下正在飞速发展进化。人类文明中任何一个全新的阶段，用时都在急剧减少，而对于人类社会的推动力，却又呈指数级增大。

而今置身于智能时代，我们每个人都将见证远比历史更为波澜壮阔的变化。

怎么读这本书？

首先要在 2020 年的具体语境中读这本书。

智博会永久落户重庆，会是重庆一年一度的永恒话题。但每一年的场景不一样。2020 年，新冠肺炎疫情给全世界制造了难题。但越是在困难面前，越要展现人类进步的力量，通过技术让世界更美好更安全。

本书记载的 2020 线上中国国际智能产业博览会（以下简称"2020 线上智博会"），在困难中创新，将线上线下、现实与虚拟进行有机结合，应用了 VR、AR、数字孪生等现代信息技术。

其次，既要读案例，又要感受书中的认知观察。

本书涉及的人工智能，既有技术发展的"来龙"，也有产业趋势的"去脉"，但从应用案例本身而言，更多聚焦在当下，将全球智能产业的浩瀚进程中，正在发生的变化，呈现在读者面前。

本书也着力将智能时代的思想认知变化穿插在其中，启发读者去进行一些深层次的思考。随着科技的深入发展，智能化的程度会越来越高，我们在享受科技带来的便利的同时，也可能承担科技带来的压力。但大势已来，未来已来，没有人能阻挡科技变革的脚步，唯有发现世界，发现自己，与时代共振。

目录

第1章
盛会：全球精英的思维演进

第 1 节　实践：从仰望星空到脚踏实地　002

第 2 节　加速：从草图线稿到泼墨挥毫　008

第 3 节　应变：人工智能置身所有变化的第一现场　015

第2章

趋势：智能产业的关键导航

第1节　抢跑5G，影响全球智能产业的关键布局　022

第2节　从新能源与碳中和，洞悉智能产业未来30年　029

第3节　产业无边界，人工智能贯穿一切场景　038

第4节　数字化生活，人工智能重新定义社会关系　045

第5节　智能红利：中国面向全球释放增长新动能　053

第3章

实践：智能产业的第一现场

第1节　智能制造，工业时代的过渡还是智能时代的萌芽？　062

第2节　智慧农业，重新定义人与农业的生产关系　071

第3节　智慧城市，城市发展史上最快的脱胎换骨　078

第4节　创新现场，人工智能变革所有行业　085

第5节　全球疫情中，中国采用人工智能实践抗疫　092

第4章

成果：科技智能的前沿创新

第1节　黑科技汇聚，从智博会眺望时代未来　102

第2节　从智能材料到自动驾驶，制造业正在加速变革　105

第3节　智博会上的新世界，生活有了"智"的飞跃　117

第4节　智慧金融变革中，科技带来"涡轮增压"效应　127

第5节　生活的便利，从智能政务开始　132

第5章

重庆：智能时代的关键见证

第1节　想象之城：重庆投身智能产业新浪潮　140

第2节　战略之城：加快建设"智造重镇""智慧名城"　147

第3节　实践之城：从名企落地到万物生长　155

第4节　未来之城：打造智能产业时代的全球智慧名城　161

后记　智能时代的年度印记　171

第 1 章

盛会：全球精英的思维演进

　　一轮史无前例的全球考验，一次义无反顾的时代征程，一座矢志不移的智慧之城，一届不见不散的产业之约。过去这一年，对于每一个国家、每一座城市、每一家企业、每一个个体，都有不同以往的意义。包围与扩散，粉碎与重塑，离别与重聚，每时每刻都在上演冲突，每时每刻也在尝试融合。

　　整个世界，曾经确定的一切仿佛都已不再确定，而智能时代的车轮仍然滚滚向前，人工智能的发展，反而成为全球不确定中最大的确定性。智博会，正是在这种情况下，在中国重庆再一次拉开帷幕。

　　当然，也有变化。从仰望星空到脚踏实地，从草图线稿到泼墨挥毫，人工智能不再局限于想象，而是以更为务实的姿态置身于所有变化的第一现场。

第 1 节　实践：
从仰望星空到脚踏实地

> 我们对真理所能表示的最大崇拜，就是要脚踏实地地去履行它。
>
> ——拉尔夫·沃尔多·爱默生

没有人能预料，2020 年会是这么突然的一年；更没有人能想到，2020 年会是这么漫长的一年。

新冠肺炎疫情给 2020 年蒙上一层阴影，几乎所有重大活动、赛事、展览、论坛与峰会都受到了影响。

这是没有体育赛事的一年，全球绝大多数重大体育赛事都推迟或取消。奥林匹克运动会是全世界规模和影响力最大的人类群体活动，自 1896 年开始每四年举办一次，100 多年来，只在两次世界大战期间中断过三次。而 2020 年东京奥运会，成为第一届延期举办的奥运会；欧洲足球五大联赛（西甲、英超、意甲、德甲、法甲）与欧洲杯，全部无限期推迟；国际篮联 U16 篮球亚锦赛和国际篮联 U16 女篮亚锦赛取消……

这是没有行业展会的一年，全球数十个国际性行业展会，纷纷被取消或延期。中国台湾台北电玩展、德国柏林亚洲服装展、中国香港巴塞尔艺术展、日本东京国际零售业应用展览会……如果一一列举出来，这个名单可以很长。

这是科技交流受限的一年，全球科技领域的多个顶级峰会被举办方彻底取消。包括世界移动通信大会（MWC）、游戏开发者大会（GDC）、谷歌 I/O 全球开发者大会、美国光纤通讯展览会及研讨会（OFC）、日本东京 LED 照明展览会（LED NEXT STAGE）……

且慢，难道在疫情困境之下，整个世界只能等待吗？

即便是在最寒冷的严冬，也有冲破冻土的嫩芽。疫情之下的冰封时刻，整个世界都需要一种继续探索与前行的力量。

微软开发者大会（Build 2020）、苹果全球开发者大会（WWDC 2020）、谷歌 Cloud Next 2020 峰会、英伟达 GPU 技术大会（GTC 2020）、Adobe Summit 2020、腾讯全球数字生态大会（2020 云上会）、2020 阿里云线上峰会、2020 百度云智峰会、华为 2020 共赢未来全球线上峰会等全球科技龙头企业的顶级峰会，最初的反应同样是延期，但后来陆续转为线上举办。

是的，世界有世界的现实难题，科技有科技的创新答案。

新冠肺炎疫情，给全球各领域产业性的展览与互动制造了障碍，然而智能产业自有智慧方式延续产业前行的脚步。

在全球科技龙头企业筹备线上活动的同时，重庆也正式向全球发出再聚智能时代的邀请函，一年一度的智能产业盛会，依旧不见不散。

2020 年 9 月 15 日，2020 线上智博会在重庆正式开幕。无论你身处全球的任何地方，都能线上逛展。2020 线上智博会突破"面对面"的地域限制，实现"屏对屏"的交流互动，开启"端

2020 线上智博会线上 VR 体验入口
The VR Entrance to the Exhibition Hall of the 2020 SCE Online

到端"的全球相聚。

在特殊的全球大环境下,重庆能够顺利举办这种高规格的全球性智能产业行业展览会,对世界各国来说,都是一种意外。新加坡人力部部长兼内政部第二部长杨莉明在开幕式上就不由感叹:"在疫情持续蔓延的当下,重庆克服重重困难与挑战举办此次盛会,充分证明了重庆的创新与数字技术应用能力。"

越是在大自然的挑战之下,越能体现全球人类命运共同体的团结。本届智博会吸引了 443 位知名专家和行业精英围绕人工智能、5G、区块链、工业互联网等前沿热点话题,碰撞思想,交流成果,举办各类论坛 41 场,551 家国内外单位线上参展发布智能产品和创新成果,集中签约重大项目 71 个,总投资 2 712 亿元,线上展厅访问量超过 1 900 万人次。

盛况不输往届,气氛尤胜从前。

第 1 章 盛会：全球精英的思维演进

智者的思想火花再一次在此碰撞，前沿的技术再一次在此交汇，新潮的应用再一次在此面世。智能产业时代全球层面的互动沟通，因新冠肺炎疫情而短暂停滞，又因 2020 线上智博会的召开而畅通无阻。

2020 线上智博会作为全球智能产业共同探索未来的新起点，再一次吹响了启程的集结号。重庆正在从城市内部发起智能化改造，大到一座智慧城市的顶层构架，小到一盏路灯的智能管理，都在融入属于智能时代的全新基因。

两江云计算中心拔地而起，给整个城市装上大脑；礼嘉智慧公园从无到有，让科幻电影里的场景照进现实；5G 信号全市重点区域全覆盖，使万物互联成为可能。

......

2020 线上智博会在重庆开幕
The opening ceremony of the 2020 SCE Online holds in Chongqing

重庆礼嘉智慧公园里，孩子们穿梭在各种"黑科技"之间，从AR乐队到全息投影，从动态水幕到智慧足球乐园，从5G自行车到智能机器人……

智博会如约而至，全球嘉宾细说别后新思考；重庆焕新相迎，智慧名城呈现智能新实践。

新技术、新业态、新模式，继续打破常规，继续叩响未来之门。

一个理想中的智能世界，在每一次全球智能产业相聚重庆、共同勾勒的线条里，墨迹由浅入深、结构攒零合整，宏伟蓝图越来越清晰。

从仰望星空到脚踏实地，从草图线稿到泼墨挥毫，重庆市连续举办了三届智博会，对全球智能产业而言，可谓草蛇灰线、伏脉千里。

工业和信息化部党组成员、副部长王志军为2020线上智博会开幕致辞
Wang Zhijun, Member of the Party Leadership Group and Vice Minister of the Ministry of Industry and Information Technology, delivers a speech at the opening of the 2020 Smart China Expo Online

第 2 节　加速：
从草图线稿到泼墨挥毫

意匠如神变化生，笔端有力任纵横。

——戴复古

回顾人类历史上所有影响广泛的创造性时刻，发端之处总是朦胧粗糙的草图线稿，繁盛之时又满是五彩纷呈的挥毫泼墨。

从最初的勾画到后来的重彩，绝非一蹴而就，其中充满了无数的启发与探索、破坏与建设、创造与重塑，置身其中，一切对美好的向往，都格外让人心动。

草图线稿的诞生

人们对未来总会有一种隐约的预感。

像是做多选题一般，先对必选项进行勾选，再逐一斟酌其他选项。

基于中国电子商务与移动互联网的发展和重庆扎实的笔记本电脑、手机制造基础，重庆选择的未来是移动通信、云计算和物联网。

不管未来是人工智能主导一切，还是机器人成为我们的忠实伴侣，或者万物通过脑电波就能读出人类的心声，移动通信是一切的基础，数据与算力都将成为新的生产资料，而物联网让设备变得更加善解人意。

2010年1月，科技部正式确认重庆市南岸区为"国家移动通信高新技术产业化基地"。2010年12月，工信部正式确认南岸区为"国家新型工业化产业示范基地"。南岸区成为国家发展移动通信和物联网的重点区域。

2011年4月6日，两江国际云计算中心暨中国国际电子商务中心重庆数据产业园在两江新区水土高新技术产业园开建，这标志着重庆信息产业从终端产品向通信领域拓展延伸，预备打造国内最大的离岸数据处理中心。

以移动通信、云计算、物联网为主干，探索新一代信息技术的发展方向和应用场景。这些规划落到草稿纸上也不过寥寥几笔。就是这简单的几笔，让重庆提前摸到了未来时代的脉搏。

成为国家智慧城市试点城市以后，草图变为线稿，智慧社会、智能政务、智慧家居、智慧商圈、智能交通等细分领域齐头并进，勾勒出一座智慧城市完整的形象。

重庆市地处中国西部，并非智能产业的核心地带，但是对拥抱面向未来的智能时代，有着更坚定不移的决心。

城市在智能化进程中加速前行，创新企业的成长也搭上了智能产业的高速列车。

借着落地巨头带来的新思维、新技术，重庆也培育出本土的人工智能独角兽企业，产业聚集效应初显，城市从内部开始新生，焕发出全新的光彩。

泼墨挥毫中，智慧城市加速布局

一座城市拥有了智能产业的草图线稿，对未来世界的蓝图，也就有了更深刻的洞察。将整个世界的智能产业创造者邀请到一起，既能为全球智能产业的画卷提供集中创作的舞台，也能为重庆智能产业的篇章沉淀浓墨重彩的实践。

在重庆举办的智博会，就是全球智能产业泼墨挥毫的第一现场。

2020线上智博会采用线上呈现、线下体验，现实与虚拟有机结合的方式，借助AR、VR、数字孪生等现代信息技术，打破了时间和空间的限制，打造全新的线上展览展示平台，平台支持展馆导览、虚拟讲解、3D产品互动、数据分析等功能，让展示内容更丰富、方式更多样。

每一届智博会，全球大数据智能化领域的最新成果，都会汇聚重庆。

2020年也不例外，硅基光技术、L4级自动驾驶中巴车、8K Micro-LED等一批新技术、新产品、新业态、新应用、新成果在渝"全球首发"。

世界的目光聚焦于此，关注重庆的未来，中国科技的未来。

2020线上智博会上，百度创始人、董事长兼首席执行官李彦宏认为，不久的将来，重庆会因为"智能化"成为广大开发者、创业者的聚集地，这得益于重庆"敢为人先"的胆识和魄力。

"数实融合"正在成为智能经济、智慧生活的发动机。随着数字世界与实体世界的融合，生产生活都在被数字化重塑。在重庆的大街上，每一条街道，每一个路灯，每一处红绿灯，都已在云端相会。住宅小区实现刷脸进小区，携带手机距离小区门禁十几米外感应解锁、小区土壤湿度低于设定值时喷灌系统自动喷水……已成为生活日常。

2020 线上智博会"智造重镇"线上场馆
Smart Manufacturing City Exhibition Hall of the 2020 SCE Online

"智慧城市"的建设浓缩成"一中心两平台"(城市大数据资源中心、数字重庆云平台、智慧城市综合服务平台),并同步建设各类"智慧名城"场景应用,以推动大数据在民用、政用、商用领域的共享互通和创新应用。

尤其是礼嘉智慧公园,全面模拟展示了重庆在智能制造、智能应用、智慧医疗、新材料、科技创新等领域的成果:陪老人打太极拳的智能机械臂、云游重庆的 5G 自行车、AR 远程手术的医疗系统、基于 5G 的远程驾驶系统……

智慧城市加速布局,智能化的成果集中涌现。

创新之下,未来已来

多年以来,重庆坚定不移走数字化、网络化、智能化创新发展之路,立足于制造业优势,已从不同角度发起对制造业的

百度创始人、董事长兼首席执行官李彦宏发表线上演讲
Robin Li, the co-founder, chairman and CEO of Baidu, delivers an online speech at the 2020 SCE Online

改造,推动制造业向高端化、智能化转型升级。

早在 2017 年 11 月,重庆市便提出要以大数据智能化引领产业转型升级,推动互联网、大数据、人工智能同实体经济深度融合,加快发展数字经济,推动制造业加速向数字化、网络化、智能化发展。

2019 年重庆市人民政府工作报告中,重点提到将培育壮大智能产业,一手抓研发创新、一手抓补链成群,首次提出要着力构建"芯屏器核网"(芯片、液晶面板、智能终端、核心零部件、物联网)全产业链,持之以恒加强基础研究,深化产学研合作,聚焦高端芯片、基础软件、核心器件等领域,加快突破一批关键核心技术，补链成群，加快智能终端产品提档升级。

2020年重庆市人民政府工作报告中，首次提出要建设"云联数算用"要素集群：统筹全市云服务资源，构建共享共用共连"一云承载"的数字重庆云平台服务体系；建设泛在互联的新一代信息网络体系，打造国际数据专用通道，实现网络体系"聚通"能力和国际信息枢纽地位显著提升；建设以数据大集中为目标的城市大数据资源中心，形成全市统一数据资源体系和数据治理架构；建设以智能中枢为核心，边缘算法、AI计算

清华大学苏世民书院院长薛澜在2020线上智博会开幕式上发表演讲

Xue Lan, Dean of Schwarzman College in Tsinghua University, delivers an speech at the opening ceremony of the 2020 SCE Online

为补充的超级算法能力,形成具备共性技术和业务协同支撑能力的算法中台;加快建设具有鲜明特色和创新引领的智能化典型应用,持续推动数字经济创新发展。

围绕丰富应用场景,将数字技术和智能元素全面融入城市规划建设管理各个领域,打造智慧社区、智能政务、智慧交通。

万物互联以后,虚拟世界与现实世界将合二为一,形成新的世界。

在未来世界,或许我们不用讲话,不用手动操作,人工智能就能通过脑电波安排好我们所需的一切。

想要光,世界就有光;渴了,水已在手边;饿了,餐食已在眼前。

时间与空间的界限,将不复存在,而世界,任由你我奔腾的思绪去改造。

第 3 节　应变：
人工智能置身所有变化的第一现场

这个伟大的世界永远旋转，不断地改变陈规。

——阿尔弗雷德·丁尼生

世界在流动中变化，每个事物都包含着自我生长的逻辑。

在当今时代，不仅是科技改变生活，更是人工智能改造一切。

从出现之初，人工智能就承载着人类对智能时代的想象，其他技术通过与人工智能融合以后，都会有质的飞跃。

智能的第一现场

未来将是"智能"的天下，世间万物都将彼此连接、交互，衍生出随心所动的生活方式。

当你还在熟睡时，轻柔音乐缓缓响起，卧室窗帘自动拉开，温暖的阳光轻洒入室，呼唤你开始新的一天。

当你起床洗漱时，营养早餐已经做好。餐毕，音响自动关机，提醒你该上班了。

辛苦一天后下班回家，车刚到车库门前，车库自动打开。

回到家，智能家居设备已经在工作，开门瞬间便亮起灯光，最适宜的室温，刚刚好的洗澡水……

这些贴心应用的背后，全是人工智能的功劳。

智能扫地机器人走遍房间的每一个角落，带走灰尘、纸屑和衣物纤维，还房间以洁净。与一般扫地机器人不同的是，智能扫地机器人能自动识别房间的情况，生成最有效率的行进动线，绕开障碍物，完成清扫。中间的每一步都不需要你我动手，一切由人工智能来代劳。

将场景换到大街上，多年以前就已经遍布摄像头。但交通事故、路面积水、电灯故障等突发事件，大多数时候仍靠人们主动反映和被动检修。人工智能加入以后，将遍布大街上的摄

惠普公司大中华区总裁庄正松在2020线上智博会开幕式上发表演讲

Chuang Zhengsong, President of HP Greater China, delivers a speech at the opening ceremony of the 2020 SCE Online

像头、路灯等事物连接成一个整体，路灯会根据光线情况开关，系统自动检测到路面积水会立即通知相关部门处理，道路拥堵则实时引导车辆分流。

人工智能无处不在，置身所有变化的第一现场。

人工智能带来的创新

人类经历了蒸汽机时代、电气化时代、信息化时代，来到智能时代，这个以互联网、物联网、大数据、云计算、机器学习、知识革命为引导的时代。

眼下热门的物联网、5G、大数据等技术，在与人工智能的融合中，激发出最大的潜力。

所有的改变背后，都有人工智能的影子。

人工智能贯穿所有应用场景，渗透生活的一切。业界对此，已有共识。

每一届智博会，都汇聚了全球创新成果，而这些成果都与人工智能有着或多或少的联系。

2020年，在新冠肺炎疫情的影响下，智慧医疗成为热门板块，市场上涌现出远程问诊、互联网医院、AR远程手术等新的看病方式。

"人工智能+医疗"在智能问诊、智能分诊、医药研发、精准医疗等多方面起到了联动作用。例如，通过语义分析可以实现病历结构化，通过智能诊断可以辅助医疗决策，通过智能化诊断设备可以推进精准医疗。更进一步，人工智能还可以在生物制药、靶向治疗等应用领域突破现有瓶颈。

机器智慧落地医疗健康行业，从电子医疗到移动医疗，逐渐转变为当前的智慧医疗，将来还会朝着具有认知能力的智能型医疗保健发展。万物互联下的智慧医疗保健，可以自动实时进行数据采集并整理分析各种接触到的环境信息。

与此同时，人工智能也成为工业互联网不可或缺的一环。

在重庆长安汽车的智能车间中，流水线上，机械臂在不断挥动。通过全连接工厂平台实现生产的人、机、料、法、环等全生产要素连接，利用人工智能的智能控制，完成智能化生产，实时管理人员、设备、产能、能耗、物流等信息，实现工厂智能化、透明化管理。

站在一个全新的时代，人们对人工智能的发问方式已经产生了根本性的变化。以前，人们会习惯性地发问："人工智能可以改变什么？"而现在，人们更希望知道："还有什么不会被人工智能改变？"

人工智能，正在改变一切。

人工智能的未来

在人工智能的引领下，技术正在快速迭代。

长安汽车两江工厂智能焊接车间
The Intelligent Welding Workshop in Liangjiang Factory of Chang'an Auto

图片来源：龙帆／视觉重庆
Photo by : Long Fan / Visual Chongqing

3D 感测摄像头、5G、人工智能云服务、AR 云、增强智能、自动驾驶、生物芯片、去中心化网络等技术，都已经脱离新兴技术炒作周期，来到高期望的峰值。

在其他技术迭代升级的同时，人工智能技术自身也在进行全方位的升级。

在 2020 年高德纳技术成熟度曲线中，人工智能依然是绝对的主角，人工智能发展潜力的覆盖范围还在不断扩大，比如复合人工智能、生成式人工智能、负责任的人工智能、人工智能增强开发、嵌入式人工智能和人工智能增强设计。

其中，生成式人工智能能够感知并动态响应不断变化的情况，为 UI / UX（可视化设计 / 可用性设计）设计人员提供实时的交互式反馈，以提高软件和智能产品的可用性。高德纳预测，生成式人工智能将用于简化数学模型和机器学习模型的创建并可随着时间的推移进行微调。

人工智能增强设计具有改变数字化和智能联网产品的设计、生产和销售方式的潜力。复合人工智能将不同的人工智能技术汇集或组合在一起，以提高学习的准确度和效率。嵌入式人工智能具有提升当前和下一代传感器准确性、洞察力和智能的潜力。生成式人工智能最常用于创建"深度伪造"视频和数字内容的技术。负责任的人工智能通过努力减少偏差来帮助企业做出更道德、更平衡的业务决策。

从 Siri 和 Alexa 等语音驱动的个人助理到自动驾驶车辆，人工智能一直根据市场的需要衍生出新的分支技术。苹果、谷歌、华为、百度、小米等一直在押注人工智能的长期增长潜力。

这些在应用过程中，衍生出来的新方向，既是对现行技术的潜力挖掘，也是对智能时代路径上的修正。随着人工智能的不断迭代升级，将会催生出新的技术、新的应用、新的生活场景，以此呼应时代的召唤。

第 2 章

趋势：智能产业的关键导航

对于风驰电掣的工业时代而言，这里是灿烂辉煌的全新高度；对于方兴未艾的智能时代而言，这里又是豪迈壮行的全新起点。不管是回望来路还是展望新程，人工智能都是时代的关键推动力，这个时代的每一寸空气，都弥漫着创新者振聋发聩的回音。

而对于整个时代的大多数人而言，登高一呼皆有回响，拔剑四顾心却茫然，是时候，为这个时代的智能产业梳理一份关键导航手册了。

第1节 抢跑5G，
影响全球智能产业的关键布局

> 科学到了最后阶段，便遇上了想象。
>
> ——维克多·雨果

2013年年底，随着工信部正式颁发4G牌照，中国与全球同步，正式进入4G时代。

当时的人们对4G的最大期待，就是网络速度能更快一点。8年过后，我们才发现，4G对整个世界的改变，远非速度那么简单，全世界的通信方式、消费方式、出行方式、金融方式、娱乐方式等生活与工作的方方面面，都已经被4G深刻地影响与改变了。

而5G到底会怎么改变世界？或许每个人都有自己的答案，但在时代面前，又不可避免地有自己的片面与局限。

理解虽然各不相同，但丝毫不影响5G成为全球智能产业最关键的布局。

在2020年的高德纳技术成熟度曲线中，5G技术逐渐成熟，已脱离新兴技术炒作周期，安然进入下一阶段。

"5G+"的预先探索

在人们的构想中,一切行业加上 5G,都能完成质变,实现产业跃迁。

在这种构想下,5G 无人矿山、5G 智能交通、5G 智慧医疗、5G 特高压变电站、5G 无人机等,都成为抢跑 5G 的新跑道。

尽管在高德纳技术成熟度曲线上,5G 技术还有 2~5 年进入成熟期,但全球的顶尖企业,早已拔足狂奔,进行各个领域的预先探索,以期提前把住智能时代的脉搏。

在 2020 线上智博会上,华为展示了软硬一体化协同发展的鲲鹏计算产业生态,以"5G+ 鲲鹏 + 昇腾 + 云"为核心,贯穿 IT 基础设施到上层应用系统全链条,在降低制造、医疗、电力、交通等领域资源使用成本方面成效良好。

------ ......

2020 线上智博会上基于"5G+ 工业互联网"打造的协同智造工厂

The Collaborative Smart Factory based on "5G+IIoT" at the 2020 SCE Online

同时，中国移动重庆公司发布"5G+工业互联网平台"，联合本地多家龙头制造企业成立5G+工业互联网实验室，打造5G工厂，深度融合5G、大数据、云计算、物联网、人工智能等新兴技术，在汽车、装备、能源、食品、医药等行业的制造生产、物流供应、设备售后等场景中，提供5G技术应用于工业制造行业的集成解决方案，探索5G在企业智能制造、远程操控和智慧工业园区建设等方面的融合创新应用。

在两路寸滩保税港空港工业园区的智慧共享物流中心，"5G+智慧物流"正在发光发热，AGV智能小车运送着货物，料箱在输送线上自动前行，堆垛机有条不紊地存货、取货。仓库外，自动驾驶汽车将取出的货物运输到附近的工厂。目前，智慧共享物流中心已建成3.8万个库位的自动化立体仓库、电子物料仓储、配送物料标识体系及智能运输体系等设施和系统。

2020线上智博会主会场
Main Venue of the 2020 SCE Online

厌倦了传统的线下购物？AR互动、云VR逛街领券、VR导购、5G直播等带你体验全新线上购物方式。重庆万象城与中国电信股份有限公司重庆分公司携手在西南地区率先落地5G+MEC商业云平台。该平台基于5G网络高带宽、低延迟的优势，可让消费者体验到丰富的AR、VR虚拟场景，并参与到娱乐互动中，足不出户即可享受沉浸式购物体验。

另外，美国高通公司分享了全球首个同时支持5G和AI的机器人平台，即高通机器人RB5平台。高通机器人RB5平台是高通专为机器人设计的高集成度整体解决方案，提供了硬件、软件和开发工具组合，能够支持开发者和厂商打造下一代具备高算力、低功耗的机器人和无人机，满足消费级、企业级、防护级、工业级和专业服务领域的要求。

2020线上智博会英国馆中，英国公司华埃莱斯集团展示了物联网全球互联平台。在这个平台上，从零售商店的电子销售终端系统到汽车行为监测，再到东非偏远农村地区的离网太阳能设施等，一切尽在网中。每隔18秒就有一个新设备、新终端或新资产连接到平台上，未来加上5G后，连接效率将会更高。

当前，以5G为代表的新技术开启了万物互联时代，新产业、新业态、新模式正加速与各行业渗透融合。

大国角逐，各有侧重

2020年对世界历史来说，是极为重要的一年。在新冠肺炎疫情的影响下，世界正在加速朝着智能时代变革。

当前，新一代网络信息技术不断创新突破，数字化、网络化、智能化深入发展，世界经济数字化转型不断加快。在新一轮科技革命中，5G技术成为智能时代变革的催化剂，人工智能核心技术突破与商业化应用落地将是关键。而人工智能的商业化之路，得靠5G来提供更快更强的支持。

各国都在加速布局 5G 标准制定、5G 基站布局、5G 应用探索等众多方面,并且形成了北美、东北亚及西欧三大主力市场。根据对频谱可用性、5G 部署进度、政府相关政策扶持与财政支持、行业企业投入情况、市场空间等多方面因素的对比,中美为第一梯队,日、韩、欧紧随其后。

目前,中国已经将 5G 引入国家战略层面,政府主导,企业攻坚,实现研发、网络建设、产业化全面推进。全国开启新基建,在 5G 网络、人工智能、工业互联网、物联网、数据中心等领域大力建设,打造新经济的基石,为数字经济发展提供强有力的支撑,引导数字经济和实体经济深度融合,推动经济高质量发展,加快数字中国建设。

同样,美国政府也在积极推动 5G 技术的发展规划,在毫米波的研究上占据领先优势。但由于毫米波发展方向受阻,缺乏丰富频谱资源,5G 可用性仍在进行方向上的探索。

2019 年 4 月 3 日,韩国的 SK 电讯、韩国电信 KT 及 LG U+ 三家电信运营商率先宣布启动 5G 服务,成为全球首个 5G 商用国家。虽然韩国对 5G 的重视程度高,但是受到自身人口的限制,市场空间较小,不足以培育 5G 的广阔应用场景。

日本的 5G 产业发展速度稍慢,在全球主要大国中排名靠后,日本三大电信公司对国内 5G 商用网络建设,也相对保守。不过,日本通信设备制造商对 6G 的研发,反而更为热衷。

在欧盟区,各成员国 5G 建设参差不齐,整体进度相对较慢,目前应用范围稍窄。

5G 具有实时在线、高并发、低延时、高可靠、高宽带、高频率的特点,将带来工业领域革命性变化,进一步拓展数字经济发展的领域和空间,为智能制造产业带来新的发展机遇。

"4G 改变生活,5G 改变社会"已是全球共识,5G 带来的不仅仅是速度上的变化,它作为万物互联的底层基础,可以叠加各种应用与创新,深度融入智能生活、智能制造、智能服务等场景,将成为全球经济增长新引擎,引发一系列的产业革命。

5G 加持下,智能产业百花齐放

对比前几代通信技术,5G 带给人类社会最大的改变,是促进了"人与人"的信息交流到"物与物"智能联动的延展。万物一旦互联,整个世界将产生奇妙的化学反应。

随着 5G 的不断普及,万物都将被连接到云端并实现交互,我们正在迈入一个由 5G 和 AI 驱动的智能云连接的新时代。

在 2020 线上智博会上,全球前沿技术,先锋应用探索,富有远见的未来预测,寥寥几笔,就勾勒出智能时代的无数种可能。

从 5G 工厂到工业互联网,从硅基光电子技术到 8K 新型显示技术,从智慧政务到智慧城市,从生猪养殖到智慧农业,各行各业都在其中找到了属于自己的位置。

R16 标准的确定、R17 的推进,为 5G 广连接的特性提供了更好的支持。我们可以通过 5G 接入更多设备,人工智能能

2020 线上智博会峰会论坛准备就绪

The summits and forums of the 2020 SCE Online are all set

够控制的设备也更多，相应地，人工智能的应用场景也更宽广。

在 5G 的加持下，智能产业已经进入高速发展阶段，各行各业进行不同路径上的实践，科技精英们各抒己见，回应着新时代的召唤。

而新时代的召唤声量，对应的是绝对的商业潜力。全球权威市场分析机构国际数据公司（IDC）预计，2020 年中国人工智能市场整体规模约为 63 亿美元，2024 年将达到 172 亿美元。

在算法理论和平台系统开发领域，欧美技术强国仍然走在世界前列，尤其在核心算法、关键设备、高端芯片、重大产品与系统等方面。

机器学习算法是人工智能的热点，开源深度学习平台是人工智能应用技术发展的核心推动力，开源框架成为国际科技巨头和独角兽布局的重点。目前，国际上广泛使用的开源框架有谷歌的 TensorFlow、微软的 DMTK 等。

中国则另辟蹊径，在语音识别、视觉识别、信息处理等核心技术领域实现了突破，具有宽阔的应用市场环境。在高德纳发布的 2020 年中国 ICT 技术成熟度曲线中，新增了一些由中国引领、在中国崛起的新技术和新业态，包括边缘计算、工作流协作、电商直播、数据中台、中台架构、云安全技术、区块链技术等。

不难看出，不论是全球企业的自我向上，还是世界各国的战略俯瞰，整个人类社会对智能产业时代的到来，已经形成共识。企业之间，国家之间，对智能产业的布局，方向虽然各有侧重，路径也是各不相同，但在 2020 年，都取得了一些关键的突破。

在全球智能化浪潮之中，5G 技术带给国家和企业新的发展机会。虽然对 5G 的最终想象还没有统一的答案，但整个世界都在朝着一致目标奔涌向前。

第 2 节　从新能源与碳中和，洞悉智能产业未来 30 年

真正的生命、真正的真理凌驾于对立的概念之上，例如金钱与信仰、机械与心灵、理性与虔诚。

——赫尔曼·黑塞

智能时代仍是想象中的未来？

不。梦想已经照进现实。

一个全新时代的发端，总是充满诸多看似对立的概念，虚拟与现实，狂热与保守，放任与克制，开放与禁忌，开发与保护。

我们需要发展数字经济，也要防范脱实就虚；我们需要探索自动驾驶，也要考虑交通安全；我们需要破译基因密码，也要担忧伦理危机；我们需要加速能源开发，也要注意环境保护。

而科技，本身就是为解决对立与冲突而生。

加速发展与减少排放，资源开发与环境保护，表面上充满了进与退、取与舍的矛盾，但本质上却是发展模式与开发方式亟待创新的深层次变革。

新冠肺炎疫情带来绿色发展新启示

整个 2020 年,全球都笼罩在新冠肺炎疫情的阴影之下,截至 2020 年 12 月 31 日,全球确诊病例总数已达 81 475 053 例,其中死亡病例 1 798 050 例。[1]

并且疫情在不同国家和地区的变异与扩散,进入 2021 年后仍然没有消退的迹象。

新冠肺炎疫情给世界经济留下持久的印记,带来永久性的变化,也给人类上了重要的一课。

未来,可能还潜伏着更大的生态危机,为了提前赢得主动权,转变发展思路,坚持绿色发展理念势在必行。

2006 年,《新牛津美国字典》将"碳中和"评为当年年度词汇,见证了日益盛行的环保文化如何"绿化"人类语言。

碳中和,也就是净零排放,指人类经济社会活动所必需的碳排放,通过森林碳汇和其他人工技术或工程手段加以捕集、利用或封存,使排放到大气中的温室气体净增量为零。

2020 年 9 月 22 日,习近平总书记在第七十五届联合国大会上提出:"中国二氧化碳排放力争于 2030 年前达到峰值,努力争取 2060 年前实现碳中和。"这不仅是中国积极应对气候变化的国策,也是基于科学论证的国家战略,既是从现实出发的行动目标,也是高瞻远瞩的长期发展战略。

能源技术的进步和创新是推动能源革命和转型发展的根本动力,也是实现"碳中和"目标的关键驱动和必然选择。

早在 2009 年,中国就推出《新能源产业振兴和发展规划》,既对太阳能、风能等可再生能源进行开发利用,也对煤化工等传统能源体系进行变革,规划期限 2009 年至 2020 年,中国在

---

[1] 中国新闻网,《世卫组织通报新冠病毒变异情况 全球新冠死亡人数超 180 万》,2021 年 1 月 1 日。

2020线上智博会"智能化应用与高品质生活高峰论坛"现场
High-end Forum on Intelligent Application and High Quality Life at the 2020 SCE Online

新能源领域的总投资超过3万亿元。[1]

从新能源到碳中和,不仅是绿色发展的理念升级,更是为智能产业未来30年的发展指明方向。

在2020线上智博会上,联合国副秘书长刘振民发言表示,发展智能技术有助于打造一个更加包容、更可持续的未来。善用智能技术,有助于消除饥饿和贫困,促进农业可持续发展,扩大教育机会,改善公共卫生条件,建设智慧城市和可持续基础设施,以及优化公共服务。智能技术将持续赋能2030年可

[1]人民网,《〈新能源产业振兴和发展规划〉择日出台》,2009年5月21日。

持续发展议程,助力实现联合国可持续发展目标。[1]

目前,中国正大力推进国内产业转型升级和高质量发展,产业结构升级能够减少碳排放、提升碳排放绩效,同时碳排放政策对产业结构升级有推动作用。

中国新能源产业新一轮投资开启,被列入多地"十四五"规划。业内预计未来10年要新增12亿千瓦以上的太阳能和风电,将带来12万亿元的巨大市场。[2]

"新能源+智能化"时代开启

人们时常喟叹,时代抛弃我们的时候,连声招呼都不打。其实不然,毕竟天明之时,打更人从不缺席,敲响的梆子声,正是新时代来临的脚步声。

从新能源到碳中和,其间的每一步,都有迹可循。

目前,全球能源革命已进入倍数阶段,"新能源+智能化"时代开启,全球的科技企业都在思考两个问题,企业自身如何实现100%的可再生能源利用?如何利用自身前沿技术驱动全社会加速碳中和?

在过去的10年,谷歌、苹果等多家国际科技企业引领100%可再生能源的潮流,已经树立了多个成功典范。

在2020线上智博会上,有众多新时代的"打更人",以及全球众多关于绿色发展的新技术、新产品。他们把新时代的讯息,传达给每一个人。

2021年,中国正式进入风电、光伏"平价上网"时代,平

---

[1] 杨野,上游新闻,《智博智见 | 联合国副秘书长刘振民:让智能技术助力实现2030可持续发展目标》,2020年9月15日。
[2] 祝嫣然,第一财经,《新能源产业新一轮投资将达12万亿,消纳问题仍是关键》,2021年1月24日。

价的绿电让互联网企业采用100%可再生能源成为可能。目前，阿里巴巴、秦淮数据、万国数据、百度的部分数据中心都已实现了较大规模的市场化绿电交易。

节约能源，从数据中心做起。腾讯滨海大厦和数据中心，通过人工智能和云计算来降低碳排放，研发的T-Block节能技术已经迭代到4.0版本。

以滨海大厦为例，8 000平方米的广场上铺装的生态陶瓷透水砖可以大量吸存和净化雨水，用来浇灌大楼里的花草；南北塔楼屋顶上的陶粒层可以达到净化雨水和减缓雨水流速、削减洪峰的效果；办公区采用的智能照明系统，每年可节电约132.61万千瓦时。

而腾讯天津数据中心，根据余热回收原理，正在研究节能应用的方案。按照这套方案，如果回收天津数据中心冬季全部的余热，将其用于采暖，覆盖的面积可达到46万平方米，可

中国海装在智博会上展示的"智慧风电远程运维管理"系统
Remote Wind Turbine Control and Maintenance Platform of CSSC displayed at the 2020 SCE Online

满足5 100多户居民的家庭用热需求。减排二氧化碳量达5.24万吨，碳排放当量约为种植286.4万棵树。

一座宜人的建筑物，不但要安全坚固，要舒适便捷且绿色节能，更要有一定的抗灾害能力。

2020线上智博会上，英国企业奥雅纳通过对建筑物数据化和信息化模型进行整合，规划出更宜居更智能的城市，通过BIM技术，协同采用3D模型，避免材料浪费并提高设计和施工质量。奥雅纳采用BIM技术设计建造了包括北京奥运场馆、广州塔、重庆来福士广场等多座智慧楼宇。另外，奥雅纳为北京首钢南区3.5平方千米的土地实施"海绵城市"策略，使新首钢高端产业综合服务区成为中国首个C40正气候发展项目，在应对雨水带来的自然灾害方面具有更好承载力。

当下，消费互联网的热浪在全球持续涌动，能源行业的产业互联网画卷也正在徐徐展开。和其他行业一样，能源行业正在从单边化走向市场化，从原始走向数字化。在能源行业向智慧化、数字化转型升级的过程中，科技企业利用云计算、大数据、物联网、5G、人工智能等高新技术赋能能源行业，助力碳中和。

其中，腾讯云推出智慧能源领域四大新品，拉开了科技企业驱动社会碳中和的序幕，包括综能工场、能源认知大脑、企业电像、智慧加油站，为能源行业打造多样化解决方案，助力能源企业的数字化转型。另外，百度、阿里巴巴、华为等头部公司亦推出了多款能源互联网相关产品。

智能产业的绿色未来

世间所有的事都一样，想要长久，就不能只执着于眼前的朝朝暮暮。没有任何一种关于人类未来的美好想象，可以建立在一个伤痕累累的地球之上。

2020软件和信息服务业高质量发展论坛现场
Forum on High-quality Development of Software and Information Service Industry 2020

———————————— ······

  这些年，全球都在推进智能产业的纵深发展，把科幻作品中的畅想变成可以触摸的现实，而未来30年，智能产业将会去往何处？

  2020线上智博会上，百度创始人、董事长兼首席执行官李彦宏直言，产业智能化的未来，将会消灭交通拥堵，提高生产、工作的效率，减少资源浪费，实现智能的便民服务，建立更加文明、安全的智能社会。

  而"智能"就是绿色发展的关键底色，智能家居、智能工厂、智能汽车、智能城市等，智能化的一切畅想，都是为了使整个世界更加绿色与高效。

  未来30年，随着智能产业的技术创新更先进、产业协同更深入、数据资源更丰富，人工智能将在能源、天气、环境、

重庆智慧气象系统充分利用人工智能技术,助力重庆防灾减灾

Taking fully advantages of AI, Chongqing Smart Meteorological System contributes to local disaster prevention and reduction

水资源等众多领域发挥巨大作用。

在能源方面,智能技术不但将驱动能源产业数字化,还会在新能源的研发和应用上,扮演不可替代的角色,催生出更加环保的能源。

在气象方面,2018年,重庆市气象局基于人工智能深度开发的"天枢""天资""知天""御天"四大系统就已亮相首届智博会,而随着人工智能前沿技术的进一步发展,这套系统也在不断升级,在日常生活、生产中,防灾减灾效果显著。

在水资源方面,中国首个全数据融合水务平台已进入建设快车道。2020年4月,阿里云与重庆水务集团宣布,合作建设重庆智慧水务,实现大数据、云计算、物联网、人工智能等新技术与水务行业的深度融合，提高处理效率、节省水资源，让水务体系变得更科学、更智慧、更绿色。

未来，以人工智能为代表的前沿科技，在产业节能减排方面的应用将成为热门，在应对地球重大挑战上的潜力也将被进一步激发。

"无绿色，不发展"，将成为新时代全新的标准。而实现真正的绿色发展，则离不开智能产业的加持。

第 3 节　产业无边界，人工智能贯穿一切场景

科学的界限就像地平线一样：你越接近它，它挪得越远。

——贝尔托·布莱希特

以前，产业与产业之间泾渭分明，大家在不同的领域里持续钻研、独立创新，隔墙互知，却鲜少来往。

经历 20 多年的互联网浪潮洗礼之后，每个领域都在变化。行业在变化，企业在变化，我们的生活也在变化。不同产业要素破墙而出，不同产业领域跨界融合，反而成为屡试不爽的全新创新方法论。

进入智能产业时代，这种跨界融合持续加速，人工智能在其中更是扮演无处不在的融合剂角色，正在擦掉产业之间最后的边界与隔阂，使所有产业都有了新的面貌，并成为流淌其中的共同血液。

人工智能无处不在

任何聚焦人工智能创新的企业，在描述自己未来的时候，都会基于想象力去构建一个拥有完美体验的智能生活场景。然

而，身处快速创新、快速变化的智能产业时代，单一科技产品总是很快进入消费者的日常生活，但这个"完美场景"却始终没有来到大众身边。毕竟，企业和产品只是构成智能生活的一个点，没有大规模技术集结和深层次技术联动的能力。

重庆在连续三年成功举办智博会的同时也在持续汇集全球的前沿技术，共同打造美好的未来世界。

在重庆两江新区礼嘉智慧公园，人工智能无处不在，黑科技云集，贯穿人们吃、穿、住、行、娱等，智慧生活触手可及。

目前，礼嘉智慧公园形成"一园五区"功能布局，打造陵江次元、云尚花林、极客社区、湖畔智芯、创新中心五大区域，一站式体验未来城市的智慧生活。

━━━━━━━━━━━━━━━ ......

机器人钢琴演奏
The robot plays the piano

你渴了，机器人可以为你制作一杯醇香的现磨咖啡；你饿了，机器人可以为你送上一碗地道的重庆小面；你困了，拥有各种健康管理定制功能的智能床可以供你小憩；你想高歌一曲时，AR 技术可以为你"创造"出一整支乐队，由你做主唱；你想看看重庆风景，5G 实时影像可以带你"一站式"游览重庆各个热门景点；你想要运动，VR 游戏可以让你打冰球、踢足球……

除了这些让我们感到新奇的生活场景，在农业、工业、服务业的众多领域，也都有人工智能的身影。

在很多人的印象中，农民总是有着黝黑的面庞、粗糙的双手，农业就是日出而作日落而息、面朝黄土背朝天的刻板工作。农业生产与人工智能，分别代表人类文明的源头与未来，除了简单的科技应用，仿佛很难有深入的交集。实际上，人工智能早已深入到现代农业的各个环节。

在智慧农业的探索中，农田四处都是感应器，温度、湿度、土壤营养度、作物生长情况等关键信息，都在智能监测系统汇成一条条数据，从而驱动一道道指令。根据指令，除草机器人在农田里穿梭，无人机在天空中喷洒农药，施肥机器人按照不同地块的营养度进行差异化施肥。农作物成熟时，收割机器人自动开始工作，农户只需拿着手机，就能知道田里发生的一切。

在智慧工业的实践中，除了科幻电影中常见的机器人以外，从工业互联网到中国智造，从智能机械臂到 5G 工厂，人工智能技术渗透其中，带动信息化与工业化的融合，推动超密集连接的物联网、车联网、工业互联网、智能制造成为可能。

在公共卫生与专业医疗领域，人工智能技术的应用也越来越深入，特别是在新冠肺炎疫情防控和疾病救治方面的应用。人工智能为智慧医院建设提供解决方案，为智慧公共卫生建设提供技术支持。各地和各级医疗机构开展智慧医院及互联网医院建设，运用预约诊疗、远程医疗等改善医疗服务。互联网医院、人工心脏、远程手术等成为热门词汇，智慧医疗指日可待。

人工智能无处不在，已在各行各业落地于众多应用场景，无人值守服务台、自动驾驶、智能门禁、云观景等，都少不了人工智能的功劳。

产业融合成必然路径

1765年，哈格里夫斯发明珍妮纺纱机，开启了工业时代的大门。之后各种新技术、新发明层出不穷并迅速应用于工业生产，内燃机的出现、电力能源的普及、现代通讯的发明……众多革命性的创新，将这个星球的居民，带入一个全新的现代社会。

在工业领域，科学和技术正有机融合在一起，人们通过科学研究来获得新的技术，技术又促进理论的提升。这个变革过程延续至今，一直在做加法。而人工智能的出现，正在让加法变成乘法，渗透更彻底，变革更颠覆。

人工智能贯穿一切产业，加速了高新技术的渗透融合与产业创新的延伸融合。农业、工业、服务业、信息产业、文化产业在同一个产业、产业链、产业网中相互渗透、相互包含、融合发展，用无形渗透有形，高端统御低端，先进提升落后，纵向带动横向，使低端产业成为高端产业的组成部分，实现产业升级。以往各个产业之间相互独立的状态被打破，形成产业融合的新趋势与新形态，这个新现象之所以形成，其背后的原因主要在于电子数据的增加、移动接口的普及和人工智能的发展。

目前，服务业与制造业互动融合成为产业融合的主要方式。比如工业设计与制造业的融合，产品要实现不断升级，就要在工业设计上下功夫，创意、技术、数据与人工智能缺一不可。

在2020线上智博会开幕前夕,2020工业互联网创新发展大会上中国移动通信集团重庆有限公司联合本地多家龙头制造企业成立的5G+工业互联网实验室正式揭牌。该实验室聚焦制造企业生产车间等实际应用场景,提供5G技术应用于工业制造行业的集成解决方案,并在此基础上进一步探索5G在企业智能制造、远程操控和智慧工业园区建设等方面的融合创新应用,不断推动传统产业的数字化转型升级。

在2020线上智博会,工业互联网领域的顶级专家、学者、企业代表齐聚一堂,共同探索工业互联网技术创新和生态构建的新路径,探讨如何加速工业互联网与实体经济融合。

2020线上智博会上,5G+工业互联网实验室项目启动仪式
The Launch Ceremony of 5G+IIoT Lab at the 2020 SCE Online

近年来，中国农村产业融合发展加快，"农业+"多业态趋势明显，鸭稻共生、中央厨房、休闲农业、智慧农业等大量涌现，用数据说话、用数据决策、用数据管理、用数据创新，不断强化数据资源"聚通用"水平。

2020线上智博会上，重庆市农委发布农业农村大数据平台，重点展示了农产品质量安全追溯和农业投入品监管平台，实现了智能监控、线上认证，通过农业生产智能化、经营网络化、管理数据化和服务在线化，促进农业三产融合。

人工智能塑造产业新体系

藩篱不再，站在时代的浪潮上，每一个人都切身感受到了世界的变化。

新的业态、新的趋势、新的概念不断推陈出新，不断触碰着人类想象力的边界。

实验证明，远在1 200千米之遥的粒子之间发生着量子纠缠，爱因斯坦称为"鬼魅似的超距作用"，速度超过光速，这预示着万物相连可能是宇宙本质。

这也是产业融合的本质，产业之间相互链接，承接产业新趋势和转移，催生新业态，比如服务业与制造业、新兴产业与传统产业、虚拟经济与实体经济、软件开发与硬件生产，形成多形式、多元化、多渠道、多层次产业融合发展的新格局。

新的时代正发生着深刻变化，产业之间的渗透融合清晰地展现出智能时代的发展图景。在不同的产业领域内，产业融合以不同的方式演进，通过科学技术的嫁接和升华，产业之间互联融合，边界逐渐模糊，无法在传统的产业分类体系中对号入座，故而演化出新的行业，最终将促成产业新体系，走向产业无边界融合。

无论业态、趋势如何变化，有一点是可以肯定的，新的产业体系非常依赖信息技术，尤其是数据与人工智能。

无边界产业由消费直接驱动，围绕不断变化的需求，需要对资源要素、产业协同和运营方式进行动态组合，消费需求的多元和变化导致这些新型产业，时刻处于一种不确定的状态，更新换代频率加快，为了满足这些需求，交叉学科的发展成为必然趋势。

而人工智能贯穿一切场景的能力，将为交叉学科的培育提供温床，也将加速产业无边界融合。

相应地，"实验室经济"也将乘风而起，以企业为主导、以实验室为载体，形成一种面向市场的技术创新模式，从研发投入到核心技术，再到产业优势，这种新型产业模式，正在成为驱动智能产业蓬勃发展的重要力量。

未来，无边界的产业生态圈将以人工智能为索引，逐渐串联起各个产业的各个环节，形成网络型不断延展的新产业链和新体系，生态群中的多个合作伙伴之间相互依赖、共生共荣，并产生协同效应，实现共同进化。

在更远的未来，或许产业唯一能确定的，就是它们正处于不断变化的过程中，随时都会拥有新的名字。

第 4 节　数字化生活，人工智能重新定义社会关系

无数人事的变化孕育在时间的胚胎里。

——威廉·莎士比亚

生活在这个时代的人们，或许没有意识到：千百年来，我们的祖先，一年又一年，一代又一代，充满无尽的重复却甚少改变，接过父辈的锄头，耕着祖辈的土地，穷其一生，未必有我们一年经历的变化大。

而我们身处的时代，正在发生的每一个今天，和以往的任何一天都不同。身处其中的每一个人，都能时刻感知这种变化的速度。

科技的发展，已经成为主宰这个世界变化的核心力量，仿佛在浩瀚的宇宙中有一支看不见的神笔，在将人类的每一个梦想、愿望绘成崭新的现实。

疫情封锁下，我们想不负春光，于是有了 VR 云赏花；相隔万里的朋友，一通视频电话，就能相见；足不出户，就能享受医疗、购物等诸多服务。

在这些数字化生活场景中，人工智能正在重塑我们的生活。

无处不在的数字化生活

从前，车马很慢，书信很远，一生能往来的朋友屈指可数，很多别离注定是永别。

现在，高铁很快，视频很近，每天能互动的朋友比比皆是，天涯海角近在咫尺。

没有任何一个时代像今天这样，人们可以同时拥有两个世界，一个线下真实的世界，一个线上虚拟的网络世界。两个世界相互融合，和谐共处，构成人们的数字化生活。

在这个时代，智能手机已经代替钱包、钥匙，成为出门必备之物。没有手机，我们就毫无安全感，通信、交通、购物、阅读、娱乐等，早已和手机合为一体，成为数字化生活的载体。

随着数字世界与实体世界的融合，生产生活都在被数字化重塑，线上向线下渗透、兼容。在人工智能技术的帮助下，不同的数字化板块链接在一起，构成人们的数字化生活。腾讯公司首席运营官任宇昕认为："未来，纯粹的'线下生活'和'传统产业'将不存在。"[1]

在数字化生活中，人们和世界沟通的方式产生巨大的变化，颠覆过往的生活方式，线上线下趋于融为一体。

腾讯的光子工作室将重庆彭水的蚩尤九黎城等苗族特色风貌植入"和平精英"、腾讯棋牌产品等游戏当中，线下景区也结合游戏IP，打造"欢乐茶馆"等特色场景，提升游客体验。"游戏+文旅"的组合，有了新的打开方式，通过联名文创产品，让年轻人爱上苗绣、苗歌、蜡染这样的传统文化，其体验式宣传每天触达上亿人次。

现实中的社区，也已经有了智慧的大脑。

---

[1] 孙磊，上游新闻·重庆商报，《腾讯任宇昕：纯粹的"线下生活"和"传统产业"将不存在》，2020年9月15日。

在重庆礼嘉智慧公园"智慧名城"展馆的智慧社区里，摄像头配有高空抛物智能检测系统，及时对高处坠落的物体进行位置溯源、动态捕捉，当发生高空抛物伤人事件时，可以整理证据链，快速找到肇事者；智慧门禁自动识别居民身份及口罩佩戴情况，符合条件才予放行；跌倒检测系统自动识别人体姿态，一旦老人在家摔倒，社区管理人员和老人亲属会立即收到报警信号，第一时间采取救助行动……而拥有这些智慧应用场景的智慧小区，两江新区已建成了 42 个。[1]

对两江新宸云顶小区的居民而言，刷脸回家、手机分享放行码、智能 APP 迅速联系物管人员，监测温度和土壤湿度自动浇灌植物，不再是实验室演示，而已经成为小区居民的日常。

在礼嘉公园的木栈道上，一只名叫"太极推手"的机械臂，正在来回推移，就像是一个太极拳师傅，人们可以在与它过招时纠正身法姿势、达到强身健体的效果。

我们的生活在人工智能的加持下，焕发出新的光彩，但太极仍是太极、景区仍是景区，只是因科技让它们有了新的打开方式，才让我们有了新的生活体验。

数字化成为全球经济的强心剂

数字产业也与线下场景融合，催生新的商业模式，拓展新的赛道。

尤其是在疫情肆虐的 2020 年，全球经济都不可逆地受到"灰犀牛"的冲击，美股熔断、原油价格暴跌，经济下行压力剧增，数字化、智能化成为众多行业的救赎契机，加强科技创新和产业合作成为不可逆转的时代潮流。

---

[1] 王倩、陈翔，重庆晨报，《一部手机畅享智慧生活 这样的智慧小区两江新区建成 42 个》，2020 年 7 月 22 日。

新加坡人力部部长兼内政部第二部长杨莉明表示,和中国等其他经济体一样,新加坡也在努力转危为机。让我们大家一起把危机转为商机,利用数字技术,发掘更大价值,创造更多商业模式,探索各行各业的新机遇。

新加坡人力部部长兼内政部第二部长杨莉明在 2020 线上智博会上发表演讲
Josephine Teo, Minister for Manpower and Second Minister for Home Affairs, Singapore, delivers an online speech at the 2020 SCE Online

在新加坡滨海湾金沙会展中心，一间混合现实演播厅中融合了增强现实、全息影像技术，尽管身在异地，也能获得亲临新加坡的体验。

通过这些前沿技术，不仅新加坡的会展行业得以重启，还拓展了新的能力，甚至重塑了一些重要领域，使得数实融合的广度不断拓展，实体数字化、数字实体化，相向而行。

未来10年，数字智能技术将从根本上改变我们所熟知的各行各业，医疗健康、工业制造、教育、金融、交通、城市管理等，而传统行业数字化将成为全球经济的新一轮增长点。

这新一轮增长中，数字化不再是大企业的专用工具，越来越多的中小企业将被数字智能技术赋能，在数字化升级中找到创新突破口，完成自身的进化。

腾讯和宗申集团旗下忽米网的合作，就是典型的数字技术赋能中小企业的案例。围绕工业互联网核心场景，为中小企业提供一站式的"新制造"升级服务。例如，重庆的太仓科技，就通过"云端工厂"实现了生产过程的透明管理，将生产周期缩短35%，原材料异常损耗降低50%，大大提升了企业效能。

随着数实融合的精度不断提高，企业的数字化升级将从台前延伸到幕后，从用户连接升级到生产环节重塑，从被迫转型发展到主动创新，从经济运行的大动脉到产业创新的毛细血管，最终形成集聚效应，实现产业数字化、城市数字化，重新定义生产方式。

人工智能重新定义社会关系

我是谁？我从哪里来？要到哪里去？

哲学终极三连问，追索的是人与世界的交互方式，是人与人的关系，人与物的关系，物与物的关系，这些的总和构成我们的社会关系。

当下,人工智能正在重新定义社会关系,数据资源成为生产资料,计算能力成为生产力,互联网成为生产关系,这些新的社会关系将带来智能时代巨大的社会变革。

近年来,人工智能发展迅速,以深度学习为例,从深度神经网络发展到循环神经网络、卷积神经网络,再到生成式对抗网络,它在不停升级换代,并触发新的创新。

在2020年高德纳技术成熟度曲线上,人工智能依旧是榜上最受瞩目的技术。而这一年的曲线上还延伸出了多个细分类别,包括复合人工智能、生成式人工智能、负责任的人工智能、人工智能增强开发、嵌入式人工智能和人工智能增强设计。

人类的历史是一部人与万物的交互史,也是追求低成本、低功耗、高效能、高便捷的持续演进史,而人工智能技术掌握着未来的交互密码。人工智能影响了关系,重新定义了关系,却不会仅限于此。人工智能的重新定义还将渗透更多的节点、场景与时空。

智能时代的交互方式到底会是什么样的?百度创始人、董事长兼首席执行官李彦宏在智博会上表示:"未来,每一个人都会拥有一个智能助手。"

百度DuerOS对话式人工智能交互系统,一直在探索人类世界与智能世界之间最佳的交互

智能时代的到来,给了这个世界无限的想象空间
The intelligent era shows unlimited possibilities to our world

方式。截至 2020 年 6 月,小度助手语音交互总次数达到了 58 亿次,越来越多的人开始熟悉并且依赖百度的人机智能交互系统。[1]

与此同时,智能社会的交互方式将降低人对手机的依赖。因为在智能产业时代,智能终端的种类与数量会远远超越手机,交互方式也会去手机中心化:从人与人之间的交互,到人与物、

---

[1]马宁宁,南方都市报,《小度科技独立融资,投后估值 200 亿元,或将加速国内上市》,2020 年 9 月 30 日。

物与物之间的交互；从有许可的交互，到无许可的交互；从单一的交互，到融合的交互。

泛在网络（Ubiquitous Network）的概念，应运而生。

关于泛在网络，专家们至今仍无法给出这个概念的精准内涵和外延，业界普遍的理解是，接入设备将无所不在，信息交换将无处不在，融合应用将无处不在。

"泛在网络"一词，超越了互联网与物联网，揭示了这个转变的实质，不再是影响和渗透，而是包围和融化。当然，人工智能大脑自身并不代表什么，当人工智能大脑、智能化的操作系统、开放共享生态组合在一起时，人工智能的场景化才真正有了泛在的力量。

在万物互联、智能互联、产业智联的泛在网络加持下，产业互联网当下业务数据化的现实需求，将进一步升级为数据业务化的未来常态、泛在连接、泛在感知、泛在交互一起重塑社会关系结构，而人工智能作为其中最关键的环节，将贯穿始终。

或许，在下下个 10 年，人们面对哲学终极三问，将会有全新的答案。

第 5 节　智能红利：
中国面向全球释放增长新动能

中国以高质量的制造著称，而且前所未有的引领创新。

——蒂姆·库克

世界来到了一个转折点上，我们已经推开智能时代的大门。新世界里究竟有什么，谁会是成为第一个挖出金矿的人，大家或许有了些许感知。

在新冠肺炎疫情的侵袭下，全球经济遭受重创。中国迅速控制疫情，利用不断升级的智能化技术推动经济复苏，成为 2020 年全球唯一实现经济正增长的主要经济体，向全球释放增长新动能。

新基建成就中国经济

一场疫情，打乱了整个世界的节奏。中国迅速地从新冠疫情的泥潭里脱身出来，回到原先的轨道。

2021 年 3 月 23 日，IMF（国际货币基金组织）发布了最新一期《世界经济展望》，报告中显示，2020 年全球经济萎缩了 3.3%，比 2008—2009 年金融危机期间的情况还要糟糕得多，仅次于二战前的"大萧条"。报告中列举了全球17个主要经济体的经济数据，中国是唯一实现经济正增长的经济体。IMF预测，中国2021年经济将增长8.4%。[1]

而中国经济增长的主要动能，除了得益于中国强有力的防疫措施和宏观经济金融政策支持，还来自于以可再生能源、电动汽车以及智能产业布局为主的新基建公共投资大幅增加。

2020年，中国首次把"新基建"写进政府工作报告，"加强新型基础设施建设，发展新一代信息网络，拓展5G应用，建设数据中心，增加充电桩、换电站等设施，推广新能源汽车，激发新消费需求、助力产业升级。"[2]

中国要持续实现经济高质量发展，构建新发展格局势在必行，中国提交给IMF的磋商报告中也明确指出，"公共投资大幅增加，有可能逆转在过去五年在实现更为平衡的增长方面取得的进展"，但毫无疑问，对新能源汽车、5G产业链、半导体产业、医药生物等领域，2020年是中国的风口之年。

而5G的布局，是新基建的重要内容。自5G商用牌照发放一年多来，中国5G网络基础设施建设稳步推进。截至2020年底，基础电信企业已建成5G基站超过60万个、5G终端连接数已突破2亿，实现全国所有地级以上城市覆盖。[3]

中国不光有前沿的智能技术，更有全球最大的消费市场和应用市场，是人工智能落地应用的首选之地。中国利用电子商务、远程办公、远程医疗、在线教育、无人技术、机器人、健康码、直播等基于新型基础设施的应用和服务，打赢了疫情防控和复工复产复苏经济"两场仗"，让人们更加直观地感受到了推动

---

[1] IMF（国际货币基金组织），《世界经济展望》，2021年4月。
[2] 新华社，《政府工作报告》，2020年5月29日。
[3] 崔爽，科技日报，《2021年我国将新建5G基站60万个》，2021年1月28日。

中国的5G"新基建"发展迅速,领跑全球
China leads the world in the development of new 5G infrastructure

新基建的重要价值和重大意义。

窥一斑而知全豹,在2020线上智博会上,重庆市政府集中展示了在新基建加持下,智能技术应用到各行各业产生的积极效应。

在2020线上智博会的智能建造体验馆中,展示了数字企业、数字项目到数字工地各层级的数字化落地应用,包含智慧劳务人脸识别系统、智慧物料验收管理、蜂鸟系统、BIM 5D数字项目解决方案、新金融等。所有这些,都可以通过硬件与软件的结合,带来沉浸式体验和多屏幕互动。

目前,重庆市数字经济发展驶入"快车道",倾力打造"智造重镇"、建设"智慧名城"。同时,因为有大数据智能化的赋能,城市管理变得越来越聪明,人民生活变得越来越多彩。

截至 2020 年 7 月底，重庆累计完成智能化改造项目 2 200 个，建成数字智能工厂 67 家、数字化车间 539 家。[1]

在广达重庆工厂车间内，一台台机器人已成为生产线上的主力。它们摇头晃脑，有的抓起零部件，准确安装在相应位置，有的举着放大镜，仔细检测产品的误差。

类似的智能制造应用，以及智慧小区、智能门禁、自动驾驶、AI 机器人等众多场景，无论是在重庆，还是在中国的其他城市，都在迅速增多。

世界拥抱"中国智造"

世界正在经历新一轮的科技革命和产业变革，中国积累的制造业和产业链优势，在智能化升级中迸发出巨大活力，物联网、智能汽车等领域蓬勃发展，智能化产品和设备已普遍进入日常生活，智慧医疗、产业互联网增长迅速，从技术跟随者到潮流引领者的角色转换，只用了短短十来年。

一直以来，美国引领着人工智能基础研究的前沿，以 DARPA（美国国防部高级研究计划局）为代表的政府机构持续推动人工智能发展与应用。

全球范围内，中美构成人工智能第一梯队；日本、英国、以色列和法国等发达国家紧跟其后，构成第二梯队。同时，在顶层设计上，多数国家强化人工智能战略布局。目前，全球包括美国、中国、欧盟、日本等近 30 个国家和地区发布了人工智能相关的战略规划和政策部署。

中国在人工智能政策上，深耕人工智能产业化，利用新一代信息技术赋能实体经济，加快"中国智造"迈向中高端，实

---

[1] 新华社，《中国西部正崛起为智能产业新高地》，2020 年 9 月 14 日。

2020线上智博会"第三届工业互联网和智能制造高峰论坛"现场

High-end Forum on the 3rd Industrial Internet and Smart Manufacturing at the 2020 SCE Online

现产业跃迁,引领新一轮世界工业和消费的潮流。

新兴技术是支持中国企业数字化转型与创新的主要引擎。即使在新冠肺炎疫情的背景下,中国科技创新风景依旧,为中国经济发展注入新动力,还催生出新经济产业,驱动着企业经营与商业模式的变化。

中国的智能化趋势,正在全球范围内引领一股新的潮流。各国政府、企业都在加强与中国的合作,依托中国新型数字基础设施,加速研发智能医疗、智能教育、智能文创、智能制造、智能农业和智能交通等诸多领域的前沿应用场景。

联合国副秘书长刘振民在2020线上智博会上表示:"未来十年,通过政府和产业间的有效合作,智能技术将持续赋能

2030年可持续发展议程，助力实现联合国可持续发展目标。"[1]

为了促进国际陆海贸易新通道沿线货物往来，提高货物通关效率，"新加坡互联贸易平台"与中国"单一窗口"的相互连通，实现两国贸易相关的数字文件无缝流转，从而降低企业成本。

在重庆，英特尔公司建立了全球最大的FPGA创新中心，面向中国、面向全世界扩展生态，汇聚人才和开发商，以期在2022年拥有10 000个FPGA的开发商工程师在线，提供在云端进行FPGA的创新，提前锁定未来。

面向全球释放增长新动能

2020年已经结束，但疫情的阴霾并未完全散去。

IMF（国际货币基金组织）对新冠肺炎疫情引发的全球衰退，做出中长期预测，认为可能带来的长期创伤效应将延续到2024年，而对中国，则给出了2021年8.4%，2022年5.6%的高增长预估。通过智能技术对传统行业的改造和升级，则被广泛评价为向全球经济释放新动能的中国智能红利。

中国政府确立了"到2025年智能制造支撑体系基本建立，重点产业初步实现智能转型"的目标[2]，并正努力提供有力度的支持，包括制定发展规划、加大财税支持、推进智能制造试点示范、加大知识产权保护等。

方兴未艾的智能化进程、不断发展的消费市场、强大的商业创新潜力等，为全球企业在中国提供了巨大发展空间。

在2020线上智博会"智造新动能·智联新产业"高峰论坛上，美国国家工程院院士、全球著名超级计算机专家陈世卿表示：

---

[1] 杨野，上游新闻，《智博智见｜联合国副秘书长刘振民：让智能技术助力实现2030可持续发展目标》。

[2] 赵宇飞、宣力祺、伍鲲鹏，新华社新媒体，《中国坚定向世界释放"智能红利"》，2020年9月18日。

当前，新型数字基础设施已成为数字经济发展的重要支撑。重庆乃至中国发展数字经济，需加快推进5G、人工智能和智能超算云平台等新型数字基础设施建设。[1]智能时代属于全世界，5G、物联网、新型数字基础设施、人工智能，也不是任何一个人、一个企业、一个国家可以独立完成的。

恩智浦半导体作为全球领先的汽车电子供应商，不断引领ADAS、车载娱乐信息系统、车载网络、汽车安全等领域的创新，在2020线上智博会上发布了S32G服务型网关处理器解决方案，标志着整车架构设计与实现的一个重要转折点。为整个服务性网关提供参考设计的，主要是重庆应用中心研发团队。

恩智浦半导体大中华区主席李廷伟表示，参考设计的开发很大程度上基于来自中国的客户需求，从这个角度上来说，我们在中国本地开发的应用已经在服务全球。[2]

而这，正是中国智能产业的"积极溢出效应"，不只提升国内智能化水平，也向外部释放智能红利，对亚洲乃至世界都至关重要，它将成为全球经济复苏的一针强心剂。

当前，战胜疫情、恢复经济是世界各国共同面对的首要任务。中国一直坚定地向全世界开放，共享全球最大的消费市场和技术应用市场，迎接智能时代的到来。

赠人玫瑰，手留余香。万物正在彼此相连，人类命运共同体和全球经济发展共同体的概念越来越深入人心。

无论现在还是未来，中国都将顺应智能化发展趋势，为世界经济复苏贡献出应有的一份力量。

---

[1]黄光红，重庆日报，《发展数字经济 重庆如何发力》，2020年9月2日。

[2]杨骏，华龙网－重庆日报，《恩智浦半导体大中华区主席李廷伟：在中国本地开发的应用已在服务全球》，2020年10月15日。

第 3 章

实践：智能产业的第一现场

　　美好想象终归要落地实践，先进的技术要被广泛应用。经历过众多技术突破、产业融合、商业对接的智能产业，终于从实验室中释放了出来，在工厂、农田、城市以及各种场景中挥洒。而其中总有一些重要的场景，成为人工智能的创新舞台。

　　出乎人们的意料，人工智能已经让工厂成为一个可以自主运行的智能制造无人区；人工智能已经让农田成为一个可以自主决策的智慧农业大管家；人工智能已经让城市成为一个可以自我进化的智慧城市生命体。

　　在这个时代，几乎所有创新之处，都有人工智能的身影，即便是在全球疫情弥漫之中，人工智能也扮演了挺身而出的实践英雄。

第 1 节　智能制造，
工业时代的过渡还是智能时代的萌芽？

科学就是对常识的不断冲击、突破和超越。

——俞吾金

第一根用于捕猎的兽骨，第一把打磨成型的石斧，第一块冶炼成型的青铜，第一台驱动火车的蒸汽机，第一根点亮灯泡的钨丝……

在人类的历史长河中，从原始石器时代的就地取材，到先进设备的发明创造，生产工具的科技水平，既是人类进化的关键标志，也是时代进步的动力来源。

进入工业时代，制造业的水平更成为衡量一个时代、一个国家、一个领域、一家企业核心竞争力的关键标尺，而从工业时代迈向智能产业时代的过程中，制造业的转型升级又成为时代变轨的关键标志。

迈入智能制造时代，人类制造业的历史也开启了全新的篇章。

工业制造与智能制造

人类文明拥有数千年的历史，而真正开启工业文明，也不过短短的两百多年。

被命名为珍妮的织布机拉开了第一次工业革命的幕布。随后的两百年里，蒸汽机、电力发动机等相继出现，机器取代手工，成为制造业的主角。

以往的工业制造要提升整体生产效率，更多依赖土地、物料、能源、设备、工具、资金和人力等有形资源要素的创造性组合。

进入智能制造时代，上述一切有形的资源要素仍然存在，但是体现制造能力的关键则更迭为数据、算力与智能协同能力等无形的资源要素。

通过智能制造技术和智能制造系统，智能制造工厂可以进行超越人工效率的生产活动，甚至在分析、推理、判断、构思和决策等方面，实现更为精确的作业，从更复杂的维度上提升整体制造能力。

正是在这种工业制造走向智能制造的产业大趋势之下，工业互联网已经成为全球制造业的全新共识。

······

2020工业互联网创新大会上，十三届全国政协经济委员会副主任、中国工业互联网推进委员会顾问刘利华致辞

Liu Lihua, Vice-Chairman of the 13th CPPCC Economic Committee and Consultant of China Industrial Internet Promotion Committee, delivers an online speech at the 2020 Industrial Internet Innovation and Development Conference

在 2020 工业互联网创新发展大会上，中国工程院院士李培根阐述了工业互联网的关键："工业互联网从系统思维、用户思维、协同思维、生态思维等四方面改变了工业企业的思维模式。未来新技术所引发的企业竞争将不再仅发生在同行业之内，同行业的竞争可能是三维甚至是更高维度的，对此，企业的生态思维要超越行业的界限。"

以前的工业机器，按照既定的程序和轨道，日复一日重复相同的程序，同时不可避免地产生需要人工校正的微小误差。

现在的智能机器，不仅体积更小、精细化操作程度更高，还拥有"智慧大脑"，能自动校正误差、提示故障并立即安排维修机器人排除故障。

在 2021 年 4 月 21 日开幕的上海车展上，有一款高性能电驱轿跑 SUV 因为"华为官宣"获得了广泛关注，成为新能源汽车中的新星，那就是"赛力斯华为智选 SF5"。这款新车由金康赛力斯与华为深度合作打造，搭载赛力斯 SEP200 电机+HUAWEI DriveONE 三合一电驱系统和华为 HiCar 车机系统，也是华为全球旗舰店开售的首款汽车产品。

身披"华为合作造车"光环的赛力斯汽车，其实是成长于重庆本土的智能电动汽车品牌，而在"百公里加速 4.68 秒的超跑级动力""NEDC1000+ 公里的续航"等硬核参数的背后，是位于重庆两江新区的智能工厂，代表智能制造的领先水准。赛力斯工厂以工业 4.0 为制造标准，以平台化、自动化、智能化、数字化为目标进行打造，拥有超 1 000 台机器人协同运作，实现了高度自动化、关键工序 100% 自动化和 24 小时在线检测，而且生产系统通过大数据和人工智能实时在线的响应方式快速精准地进行 C2M 规模定制生产。[1]

---

[1] 张婧，华龙网，《上海车展的火爆延续 赛力斯华为智选 SF5 为何这么红》，2021 年 5 月 29 日。

赛力斯汽车智能工厂的关键工序实现了100%自动化
The critical process of SERES smart factory is 100% automated

重庆七腾科技公司自主研发的SUPPORT P2消防灭火侦察机器人，是真正的"烈火英雄"，该款机器人拥有强大的灭火功能，拥有80升/秒最大流量强力水炮，能轻松拖动12根120米充满水的80水带，实现火灾现场迅速增援，尤其适用于石化、燃气等易爆环境，对提高救援安全性、降低消防人员风险具有重要意义。[1]

通过应用智能机器和系统，不仅能制造出工业品和生活消费品，更带给了人们一种全新的解决方案。

---

[1] 刘磊，新华网，《长寿：吉祥物"柚柚"带你"云游"2020线上智博会》，2020年9月14日。

智能制造的质变在于互联

根据六度分离理论,每个人平均只需要通过 6 个中间人,就能与全世界任何一个人建立联系,不管对方在哪个国家,是哪种肤色。

人与人之间的联系,尽管可以被理论定义,但在现实中却难以将其运用到极致。

反观智能制造业,则正在现实世界创造一个更为宏大的真实互联过程。中国航空工业集团信息技术中心原首席顾问宁振波,在 2020 线上智博会上分享了一组数据:工业互联网的发展,人、机、物的互联互通是最重要的,全球目前有工业总线约 120 多种,通信协议超过 5 000 种。

海量工业互联,正以超出人们想象的速度,实现人与机器、机器与机器、机器与信息的连接。然而,智能制造的优势并不止于此,机器学习的即时进化,正在刷新人们对智能制造的认知。

2020 线上智博会上,图灵奖得主大卫·帕特森对此表达了自己的观点:"机器学习的第一个词是'机器',我们需要速度更快的机器及大量的数据。而多亏有了物联网,我们才能获得大量数据。云计算又帮助我们将数据集合到一起。"[1] 在机器智能的基础上,基于互联,基于智能,万物皆可连接,还有更广阔的应用空间。

三菱电机将人脸识别技术应用在了电梯上,开发了"ICS 智能呼梯系统"。该系统包含了手机 APP 呼梯、人脸识别呼梯两大非接触识别的呼梯方式。乘客可使用针对乘梯用户推出的手机 APP 中的蓝牙呼梯功能召唤电梯。在办公场景中,配备人脸识别技术的 DOAS 目的层预报系统,通过预分配来避免轿厢

---

[1] 杨野,上游新闻 - 重庆晨报,《图灵奖获得者大卫·帕特森:我们需要更快的机器来推动人工智能发展》,2020 年 9 月 15 日。

拥挤,并最大限度减少乘梯者的候梯时间、乘梯时间;而在住宅电梯场景中,业主也可通过配备人脸识别功能的轿内操纵箱登记楼层。

目前,智能智造已经成为新时代的重要主题之一,而后续的运维服务也有了新的打开方式。在2020线上智博会体验展区,中国联通运用5G+AR技术,模拟汽车远程运维服务场景,人们通过佩戴AR眼镜即可现场连线汽车远程服务专家,并基于第一视角画面进行音视频通讯,实时解决用车过程中的突发问题。同时,通过5G终端扫描汽车方向盘,3D呈现汽车使用说明书,为人们带来全新用车体验。

智能制造在强调资源整合和价值增值的改变中,指向产业系统性变化,引发新模式、新业态的出现。

通过5G+AR远程运维服务,可实时解决用车过程中的突发问题

5G+AR remote control and maintenance service handle the accidents happened in driving in real time

智能时代快速绽放的花苞

智能制造,到底是工业时代的过渡,还是智能时代的萌芽?

这个问题现在已经不难回答,智能制造与智能互联早已通过人工智能、物联网、5G等技术衔接到一起,孕育出智能时代的花苞。

通过新一代信息通信技术与先进制造技术深度融合,设计、生产、管理、服务等制造业各个环节有了自感知、自学习、自决策、自执行、自适应等功能,从最初的生产制造为主向全生命周期延伸,直至演化出像社会化制造、云制造、泛在制造、物联制造、信息物理系统制造等一批具有未来增长潜力的智能制造模式。

———————————————— ······

> 未来已来,智能科技正在连接一切
> Intelligent technologies connect everything and make the "future" become reality

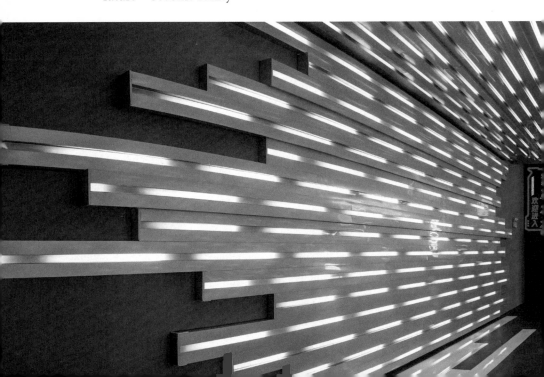

第 3 章　实践：智能产业的第一现场

人工智能从 1956 年开始萌芽，历经 50 多年，到 2008 年，人工智能已经在不同国家被赋予了全新的战略使命。尤其是以美国、中国、德国为首的制造业强国，先后提出"美国先进制造""中国制造 2025""德国工业 4.0"等国家战略。

目前，已有越来越多的制造企业加入智能工厂、数字化工厂作为智能制造重要的实践领域，融合移动通信网络、数据传感监测、信息交互集成、高级人工智能等智能制造相关技术、产品及系统，在工厂层面进行具体应用，以实现生产系统的智能化、网络化、柔性化、绿色化。

如海尔与美的等创新先行者，已经成为各个行业建设智能工厂的参照样板。

新一代信息技术与经济社会深度交融，以 5G 等新一代信息技术为代表的新一轮科技革命和产业变革正在快速孕育兴起，其速度、广度、深度前所未有，5G 作为激发物联网潜力

的催化剂,成为新一代智能工厂的"香饽饽"。

在 2020 线上智博会上,中国联通联合长安汽车通过工业沙盘现场展示"5G+工业互联网"协同智造工厂生产全过程。利用 5G、边缘云计算、工业互联网等新一代信息技术与汽车生产制造的融合,全面展现数字化、无人化、网络化技术在生产车间的综合应用,打造了汽车行业首个基于"5G+工业互联网"的协同智造工厂。

在新一代信息技术和先进制造技术推动下,制造企业利用单元、系统和管理组织等方面创新,优化生产过程,提升产品服务价值,链接更多应用场景。

我们期望的智能生活是万物可以互联,人工智能渗透一切,生活中所需的服务都能由人工智能精准地洞察并即时提供。不难看出,智能制造已经把曾经的梦想照进了现实。

2020 年 8 月 11 日,雷军在小米 10 周年演讲上首次揭秘被誉为"黑灯工厂"的小米智能工厂。工厂内部采用全自动生产线,甚至不用开灯,无人干预,一年就能产出百万台手机。而在规划中的"黑灯工厂"二期,一个工厂可能只有一百名工程师,却可能实现一年六七百亿产值。

稍不注意,智能制造这朵时代之苞,已经出人意料地在全球快速绽放。

第 2 节　智慧农业，重新定义人与农业的生产关系

此刻的一切完美事物，无一不是创新的结果。

——约翰·穆勒

农业是人类文明的起点。但是一提到农业，我们的脑海中往往充斥着各种刻板的印象，比如起早贪黑的辛劳、入不敷出的收益。

然而依托于智能产业的发展，智慧农业已经开始焕发出新的生机，人们通过人工智能技术与农业生产相结合，已经实现了无人化、自动化、智能化管理。

时至今日，融入新一代信息技术的智慧农业，面目日渐清晰，大数据、农业机器人、远程遥感技术等早已照进现实，农业 4.0 时代近在眼前。

智博会上的农业黑科技

2021 年 2 月 25 日，习近平总书记在全国脱贫攻坚总结表彰大会上庄严宣告："经过全党全国各族人民共同努力，在迎来中国共产党成立一百周年的重要时刻，我国脱贫攻坚战取得

在2020线上智博会上,重庆市各区县展示大数据智能化发展成果

The districts in Chongqing showcase their achievements in intellectualized development of big data at the 2020 SCE Online

……

了全面胜利,现行标准下9 899万农村贫困人口全部脱贫,832个贫困县全部摘帽,12.8万个贫困村全部出列,区域性整体贫困得到解决,完成了消除绝对贫困的艰巨任务,创造了又一个彪炳史册的人间奇迹!这是中国人民的伟大光荣,是中国共产党的伟大光荣,是中华民族的伟大光荣!"宣布这个好消息后,中国在第一时间收到了来自国际社会的诸多祝贺。其实国际媒体对中国的脱贫攻坚事业,也是关注已久,早在2020年9月,中国对外宣布已成功使8亿多人摆脱贫困的成绩,外媒报道中,不吝"人类历史上的一大壮举"之类的溢美之词。美国哥伦比亚大学可持续发展研究中心主任杰弗里·萨克斯则说道:"中国与贫困的斗争一直是人类历史上最杰出的斗争之一,并且对

世界其他地区提供了借鉴。"[1]

脱贫攻坚胜利的背后,是一个政党全面消除贫困的坚定决心,也是一个国家全力拥抱科技的强烈信念。

农业是根植中国人血脉的生命基因,14亿中国人口中的9亿已经通过快速城市化进程摆脱了土地的束缚,而一个转身,他们又携带着科技的力量,转身为故土乡亲探索出了一条科技脱贫之路。

在2020线上智博会上,众多农业黑科技仍在不断涌现。依托农业基础设施、各种传感器、通信技术,实现全程感知、预警、分析、决策、执行、控制的智能化,使农业生产更"智慧"。

在重庆梁平区金带街道,占地400亩的数谷农场里,

---

[1] 曲颂、王莉、黄培昭、姜波、殷新宇、谭武军、方莹馨、刘旭霞、丁子,人民日报,《"人类历史上的一大壮举"(外媒看中国)》,2020年9月29日。

35 000平方米的智能温室中，水肥一体化灌溉技术、荷兰番茄轨道种植技术、仿生立体式栽培、螺旋仿生栽培、多层管道栽培、摇摆式追光栽培、垂钓式气雾栽培、算珠式栽培等智慧农业栽培模式应有尽有。

智联总控中心是农场的心脏，管理平台集园区管理、四情监测、质量全程追溯、生产操作运用、农业大数据等模块于一体，可实现农业生产经营的土壤监测、病虫害监测、安全溯源、可视化、自动化、智慧化。

通过这个服务平台，可在任何有网络的地方实时观察农作物生长状况、操控温室设备设施、指导园区生产作业，并能设置农作物生长环境的自动运行模式，确保农作物生长所需的最佳环境。实现从种子到餐桌的全过程质量追溯，可有效控制各个环节的质量安全隐患，为产品高质量提供保证。

和传统农业相比，信息和知识已成为智慧农业生产最重要的元素，运用于生产过程从而形成知识生产力。智能化的装备将逐渐参与到农业作业的全过程，实现机器换人，减轻体力劳动。

如今，随着5G新技术应用到农业领域，田间地头早已开出"智慧的花朵"。

在重庆市农业科学院科研基地，一架5G网联植保无人机，可为试验农田提供集成无人机植保、遥感大数据和农业大数据于一体的高效飞防服务和精准农业服务。同时，5G网联植保无人机数据还将对接重庆农科院智慧农业园区数字化管理平台，实时监测园区内不同产业区土壤、环境、病虫害等田间信息，实现农作物生长的动态评估等，高效科学指导农事生产。

目前，重庆已建成农业农村大数据平台，包括数据标准和平台运行两个体系、一个农业大数据资源库、政务服务和公众服务两个系统，初步形成了农村农业大数据应用成果矩阵，推动了大数据在现代农业领域的创新应用。比如，"渝益农"全市信息进村入户大数据平台，全面集成公益、便民、电商和培训服务，促进信息服务延伸到村、信息精准到户。

同时，农业农村大数据平台还促进了农产品价格监测预警。受疫情影响，脐橙传统销售渠道受到冲击，价格一路走低，倚靠平台的监测预警，奉节脐橙调整营销策略，避开了全国脐橙市场价格持续低迷的阶段，通过差异化渠道、数字化营销，开展全网直播卖货、防疫物流一卡通等创新方式，在疫情防控最严格的2020年2月实现了日销3 000吨的佳绩。

丰都县丰泽园肥业有限公司化验室，化验员对产品的养分含量和有机质进行检测

The analyst tests the nutrient contents and organic matter for products in Lab of Fengzeyuan Fertilizer Co., Ltd in Fengdu county, Chongqing

图片来源：崔力／视觉重庆
Photo by: Cui Li / Visual Chongqing

智慧农业 4.0，重新定义农业

清晨第一缕阳光照亮大地，农田里的灌溉系统就已经开始"上班打卡"，田里的感应器也进入了"会议模式"，互相交流土壤的温度、湿度、营养情况，汇总数据并很快安排好当天的任务，"散会"后，施肥机器人自动给营养不足的地块施肥，而播撒农药的无人机逐步飞遍农田的每一个角落……

在这个过程中，从一颗种子的挑选到播种、生长到收获、加工到销售，每一步都在大数据中，可以看见，可以追溯。

智慧农业 4.0，各种人工智能加持的智能农机，不但可以取代农户繁杂辛劳的体力劳动，而且可成为农户的智慧大脑，提升农业生产的整体经济效益。

传统农业生产和运作依靠经验，农作物的产出依赖于个人努力和自然条件。当人类社会不断进步，技术革命不断出现，传统农业也从过去的自给自足、手工劳作的原始模式逐步向现代农业进化，不断启用新的智能系统与机械技术。

由此，个人的努力与劳动，不再是决定农业产能的重大要素。人从与农业强绑定的关系中剥离开来，成为生产关系中的客体。

从另一方面来说，农业也从个人的局限性中解放出来，不再完全依赖祖辈传承的经验。人工智能、大数据、物联网等都是人类集智的结果，农学专家通过整合湿度、土壤质量、空气指标、天气预测等相关历史数据，对农作物与其他相关因素的数据加以分析，找到种植农产品的最佳配比，从而大幅度提高农产品的产量和质量。

未来，在大数据运作成熟时期，食品安全的监管部门与机构也可以通过在线数据库、互联网、移动智能终端和社交媒体等手段进行食品安全的数据收集。特别是进入到 5G 时代以后，智能手机将有望具备食品感应监测功能,并将生产记录同步到计算机和官方食品数据中心,从而将食品安全置于全民监督之下,形成良性监管。

"科技兴农"从来都不是一句简单的口号。只不过,假以时日,这句话可以改一改了。当大数据与农业相结合,现代农业将再次升级,万物互联和大数据应用于现代化的农业当中,将为我们展现出一幅优美的画卷,人与农业的关系再一次被改写,"科技兴农"也将改为"智慧兴农"。

如忒修斯之船一般,在更换完所有的旧船板以后,它已脱胎换骨,虽然航行的方向与目的没有变更,但它可能会航行得更快、更精准。它既是当初那艘忒修斯之船,却又不再是原来的忒修斯之船。

农业也一样,在智能产业时代科技创新的锻造下,每一个环节都焕然一新,却初心不改,步履更健。

第 3 节　智慧城市，
城市发展史上最快的脱胎换骨

> 对于一个城市来说，最重要的不是建筑，而是规划。
>
> ——贝聿铭

我们终会老去，但城市可以永远年轻。

城市是人类社会发展到一定历史阶段的产物，是文明产生的标志之一。

智能时代的来临，让一部分主动拥抱智能产业的城市有了全新的发展契机，新兴城市快速崛起，老牌城市改头换面，书写城市发展史上最快的脱胎换骨。

新兴智慧城市崭露头角

智能时代的浪潮席卷全球，人工智能的产业星火撒遍世界的每一个角落。

有的城市以创新技术为导向，有的城市以产业聚集为目标，有的城市以智慧应用为突破，不同城市在智能时代的路径选择上各有不同，但殊途同归，都会是智能时代的城市新星。

重庆积极参与智慧城市实践，重庆市南岸区、两江新区被列为中国首批国家智慧城市试点区域。经过多年的摸索和建设,如今的重庆已跻身全球智慧城市,在民生服务、城市治理、政府管理、产业融合、生态宜居五个方面取得了阶段性成果。

在重庆智慧城市建设中,通过推动全市数据大集中、大融合,并从法规标准、开放平台、创新应用、数据要素市场培育等方面建立健全了数据开放服务体系。特别是"云长制"实施一年多以来,"管云管数管用"取得较好效果,截至2020年12月全市累计数据调用量增长168.6%。[1]

2020线上智博会"智慧名城"展厅
The "Smart City" Exhibition Hall of the 2020 SCE Online

---

[1] 夏元,重庆日报,《重庆获评"2020中国领军智慧城市"》,2020年12月15日。

目前，重庆已建成智慧城管大数据中心，并持续加大对物联网应用试点的建设，成为全国第一家省级（直辖市）层面城市管理行业大数据中心，城市照明智能控制系统建成率已经达到了90%，已汇集数据3 400万条、视频监控2万多路，城市管理初步实现"设施监控、人车监管、惠民服务"一图呈现、一屏通览、一网统管。

在"2020线上中国国际智能产业博览会——智能化应用与高品质生活高峰论坛新闻发布会"上，有专家和领导特别分享了重庆打造的同上一堂课、最多付一次、占道自动报、全天候监护、热力预警图等17个"小切口、大民生"便民利民的智能化创新应用试点项目，涉及线上教育、线上支付、线上管理、线上养老、线上出行等领域。为人们的生活带来新的打开方式。

一座城市智能化水平的提升，会使置身其中的市民，工作与生活更加智慧便捷。重庆市线上政府管理平台"渝快办"的用户已经突破了1 200万人，移动端上线服务的事项已经达到了1 016项，新上线的区县特色服务事项超过了400项。[1] 按照重庆的城市规划，未来将按照分级分类推动全市智能化应用创新发展，尽快实现智慧生活全民共享、城市治理全网覆盖、政务协同全渝通办、生态宜居全域美丽、产业提质全面融合、基础设施全城连接。

重庆正在将"智造重镇"和"智慧名城"打造成自己的新城市"名片"，也为全球智慧城市的发展提供了一个参考。人工智能、大数据、区块链、工业互联网等信息技术从根本上改

---

[1] 重庆市人民政府新闻办公室，《重庆举行2020线上中国国际智能产业博览会智能化应用与高品质生活高峰论坛新闻发布会》，2020年9月1日。

变了企业运营、人民生活、公共管理方式，在"数字城市""信息城市""无线城市""泛在城市"和"智能城市"等维度上形成了新的拓展。

近几年，韩国松岛新城在全球声名鹊起，这座2015年才完成填海造城的人工岛屿，从一开始就被设计成一座"数字之城"。整个城市的社区、医院、公司、政府机构实现全方位的信息共享，数字技术深入到每一户。居民使用智能卡就能完成大部分生活应用，包括支付、查账、寻车、开门等。虽然现在看来智能卡技术已有些过时，但数字技术的深入是实实在在的，把智能卡换成手机APP，服务依旧成型。

除此以外，巴西里约热内卢的智慧市政中心引领智慧政务的潮流，冰岛雷克雅未克在能源可持续性和智慧解决方案领域处于世界领先地位，丹麦哥本哈根在环境智慧治理方面表现突出，新兴智慧城市如雨后春笋般在全球崭露头角。

老牌城市有了新名片

在新兴智慧城市屹立潮头之时，全球知名的老牌城市也在乘风破浪，在原有的城市基础上，实施智能化改造，旧貌换新颜，有了新名片。

本质上，智慧城市是利用信息和通信技术对一个地区或城市进行再开发，以提高城市的性能和服务质量，如能源、交通、公用事业和互联互通能力，从而提高市民的生活质量。

一直以来，新加坡作为全球重要的金融中心、航运中心和贸易中心而被世人所熟知。然而鲜为人知的是，多项报告和研究都显示，新加坡智慧城市建设水平在全球名列前茅。

推动市政服务数字化升级，是新加坡建设智慧城市发展重点之一。用户只需在该系统进行一次全国数字身份认证，就可获得60多个政府机构的在线服务，包括查询公积金存额和申请租屋等。

在新加坡，已经实现光纤到户全覆盖，并在此基础上致力于打造一个全国传感网，成千上万的传感器分散在城市各处。新加坡要发展为"智慧国"，甚至厕所也开始走高科技路线，一旦厕所发出异味，监测系统便会自动提醒清洁工人前去清洗，确保厕所时刻干净。

美国硅谷，是世界最重要的技术研发基地之一，而旧金山"近水楼台先得月"，多年来，一直是全球领先的智慧城市之一，致力于利用新一代信息技术使建筑运营更加高效，减少能源使用，简化废物管理系统，改善运输系统。

几年前，旧金山就已经用更环保的智能LED替换了18 500个轻压钠路灯灯具，灯具内置的无线智能控制器，不但具有远程监视灯光的功能，而且能在每盏灯烧坏时发出警告，从而提高安全性并节省维护成本。

旧金山最广为人知的城市标签，是美国西海岸最重要的金

重庆两江新区正以礼嘉智慧公园、两江数字经济产业园等重点平台为引领，打造一座智慧之城
Chongqing Liangjiang New Area strives to build Chongqing into a smart city led by key platforms as Lijia Smart Park and Liangjiang New Area Digital Economy Industrial Park

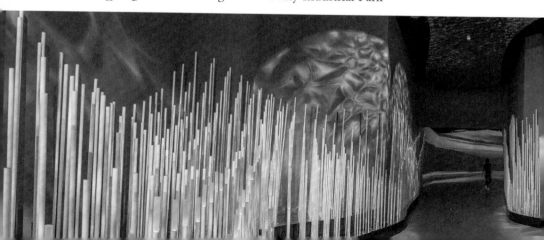

融中心及联合国的诞生地,自从推动智慧城市战略以来,又多了一张新的名片——北美最环保城市。

英国首都伦敦是世界金融中心,也是全世界博物馆、图书馆和体育馆数量最多的城市,有着丰厚的历史底蕴,也是一座拥有最多初创企业和程序员的城市。

从 2009 年发布的"数字英国"再到 2013 年的《智慧伦敦计划》,伦敦不断朝着智能化方向前进。从智慧出行到智慧政务,再到智慧社区,伦敦都进行了全方位的智能化升级。

目前,伦敦金融城已经设置遍布全市的带有液晶显示屏的数字化垃圾回收箱,所有垃圾回收箱连接 Wi-Fi 信号,指导居民对垃圾分类处理,同时可以收取天气、时间以及股市行情动态等信息,有效助推伦敦智慧城市建设。

智慧城市畅想

"好城市"到底长什么样子?莎士比亚回答说:"城市即人"。

不论科技发展到什么程度,智慧城市建设应以人为本,最终落脚点是为城市中的人们创造更美好的生活。在智慧城市的发展进程中,终极目标是智慧社会与智能时代,城市只是智能时代的载体,是微缩的实践样本。

尤其是新冠肺炎疫情之下，人们对智能化的城市生活方案也充满了越来越多的想象和期待。

通过远程操控技术、5G、大数据、人工智能等前沿科技技术，可以极大地帮助疫情期间不便出门的居民、企业员工解决生活及工作的难题，智慧城市的建设仍在加码。人们生活其中，只需一个按钮或者一个信号指令，就可以随心所欲享受到各种服务，幸福感、舒适度和满意度都接近满分。

城市的每一个角落，每一个物品，相互连接，有条不紊地履行着自己的职责。

智能路灯根据天光明暗调节亮度，红绿灯根据行人和车流给出最有效率的信号切换，智慧回收车在半夜自动带走城市所有的垃圾；天刚微亮，洒水机器人已经为城市的绿荫送去清凉；大街清扫、交通巡逻都由机器人完成，城市的天气预报精确到秒……

城市的出现，是人类走向成熟和文明的标志，也是人类群居生活的高级形式。经历数千年的发展，人类社会早已升级了"筑城以防卫、建市以交易"的原始城市概念，为城市不断注入新的内涵。

而进入智能产业时代，智慧城市的出现，正在让更多的城市脱胎换骨，但无论城市怎样进化，"人"仍然是城市的主人，就像 2020 年 7 月 IESE 发布《2020 年城市动态指数》（第七版）得出的一系列结论与建议，排在第一条的仍然是："以人为本"。

第 4 节 创新现场，
人工智能变革所有行业

每天都会出现一些新的奇迹，戏言变成了真实。

——塞万提斯·萨维德拉

没有驾驶员，车、船照样能安全行驶；走进银行，只需刷一下脸，就能获得量身定制的理财方案；入住酒店，只需动动口，就能开关窗帘和电灯，调节空调温度；街道垃圾桶满了会自动上报，及时得到清理……

这些曾经的幻想，都已变成现实。

而幻想成真的前提，就是人工智能悄然渗透到世界上的各个角落。

生活中的"黑科技"

"人工智能将渗透一切！"

这早已不是什么新鲜的观点，早在两年前，高德纳公司发布 2019 年技术成熟度曲线之时，就已点明这个趋势。

经过两年的钻研和发展，人工智能早已渗透进各行各业，新冠肺炎疫情更是加速了这一进程。车、船、酒店、机场、银

图片来源：刘松林 / 视觉重庆
Photo by : Liu Songlin / Visual Chongqing

人工智能为重庆武隆仙女山的游客提供了数字化新体验
Artificial intelligence in Chongqing Wulong Fairy Mountain digitalizes the sightseeing experience for tourists

行、景区、小区、农贸市场、公厕等很多地方，突然变得有"智慧"了。

现在出去旅游，不再需要事先查找攻略、安排路线和食宿。人工智能催生出众多文旅黑科技，比如在重庆武隆仙女山景区，可以通过"一部手机游武隆"这个微信小程序，使用导览助手、线路推荐、门票预订等多项功能。

到了武隆仙女山景区，点击"慧玩武隆"栏目，即可根据游客当前所在位置，自动推荐附近美食的位置；在出游前，点击"武隆新玩法"栏目，可在手机上完成出游路线定制。其中，导览助手不仅有语音讲解，还结合实时旅游场景串连武隆周边"吃住行游购娱"旅游消费资源，为游客提供实时引导服务。而这些服务由人工智能串联起来，实现景区"智慧"游览。

科大讯飞全球首个 AI 多语种虚拟主播小晴，也"现身"2020线上智博会并担纲主持，生动展示了科大讯飞人工智能语音技术的实力与速度。同时，科大讯飞在会上发布了最新的产品。讯飞智能学习机，带来人工智能对教育的重新变革；展出的讯飞听见、讯飞智能办公本等人工智能科技产品，在政法、医疗、生活服务等众多领域落地应用，从不同方面探索人工智能发展的宽度。

在中国多个城市已经开通的实验路段，涌现全程由人工智能掌控的自动驾驶出租车。

从 2020 年到 2021 年，国内的智能汽车创新厂商们已经发布了多款互联网汽车，汽车产品的属性正在发生变化，由一种配置电子设备的机械装置，转变成一种配置了机械装置的电子设备。

在购物场景中，基于人工智能的个性化定制来了。人工智能服装定制平台，通过三维扫描仪，扫描和生成特定姿势的人

智慧文旅，科技与旅游融合下的交互式体验
Smart Tourism—smart and digite cultural tourism brings the interactive experience by integrating technology and tourism

体模型，通过人体重构技术为扫描人体自动加入骨骼，使其能够虚拟穿装和展示，让消费者在虚拟空间看到自己穿衣的效果，实现全方位的体验式消费，实现服装个性化定制，轻松一键换衣。

而人工智能的渗透不光在我们的吃穿住行上，甚至连最不起眼的公共厕所都已有了新面貌。重庆智通云厕公司建设的智慧公厕，通过人工智能、红外线感知、环境感知和无线传输等技术，让每间厕所的温度、湿度、气味浓度和蹲位使用情况一目了然。智慧公厕除了给使用者带来智能体验，同时也在践行环保理念，配套的人脸识别供纸机，需要使用者人脸识别才能取纸，而且每次只会提供一张70厘米长的手纸，如果需要再一次取纸，必须间隔十分钟，或者采取手机扫描二维码，这样不但可以让大多数人节约用纸，也能兼顾特殊场景下的用纸需求。

人工智能的应用，已经为人类开启一个新的时代。

生产中的"新动能"

人们在生活中的感知有限，而人工智能技术更多地藏在工业生产和城市建设的背后，从源头上改变着人们的生活。拥抱智能时代变革、促进高质量发展、提升业务韧性，是企业成长与发展的全新机会。

其中，川开电气的智慧工厂建设为制造企业的智能化转型提供了参考。施耐德电气利用人工智能技术为川开电气量身定制了从精益咨询到落地实施的一体化解决方案，重新打造三条精益化生产线，实现了精细灵活的排产方式和物料合理配送，降低了70%的在制品数量。

改造过后，不但减少了生产过程中的七大浪费，使生产效率有效提升16%，并且大幅度提升了项目交付和供应链的能力，有效提高了仓库利用率，降低了运营成本。在施耐德电气的助力下，川开电气打破电气行业壁垒，让数字化真正落实到生产和经营中，实现了成功转型，并构建出一条差异化的竞争新路径。

智能制造的智慧化，不仅是工厂自己的事情，也正与所处的智慧城市产生更紧密的联动。

中国的人工智能开发者群体需要更丰富、更强大的人工智能基础设施，包括新型基础设施、新型 AI 芯片、便捷高效的云服务及各种应用开发平台、开放的深度学习框架、通用人工智能算法等。

百度的飞桨深度学习平台，是中国自主研发的第一个深度学习框架，造就"智能时代的操作系统"。从 2016 年开源以来，飞桨平台让各行企业、开发者都能借助它来开发人工智能应用，截至 2021 年 5 月 20 日，飞桨开发者数量累计达到 320 万，相比一年前增长 70%。[1]

可以说，深度学习平台已经推动了更丰富的人工智能创新，广泛应用到各行各业，让智慧城市成为可能。

2020 年 8 月，重庆市新型智慧城市运行管理中心建成投用，这是智慧城市的智能中枢，旗下的数据资源中心汇聚重庆各类政务数据资源，已接入重庆市政府办公厅、市生态环境局、市卫生健康委、市城管局等 21 个部门、区县和单位共计 43 个系统，涉及线上管理、线上服务、线上业态、"小切口、大民生"等不同领域应用系统。[2]

智慧中枢不但借助监测预警中心深度挖掘数据价值，实现对城市运行态势感知、监测分析和预测预警，还能调度指挥中心实现对日常工作监督调度、城市突发事件应急处置调度，而

---

[1] 林志佳，钛媒体，《百度 CTO 王海峰：百度飞桨平台开发者数量达 320 万，同比增长 70%》，2021 年 5 月 20 日。

[2] 黄兴，新华社，《重庆建成投用智慧城市"智能中枢"》，2020 年 8 月 22 日。

且可综合赋能平台提供共性技术、业务协同、安全运行等服务支撑保障,让城市的"大脑"最终实现一网统管、一网通办、一网调度、一网治理。

后疫情时代的人工智能

后疫情时代,全球都面临着巨大的不确定性。

好在借助新一代信息技术的力量,在全球数字化、智能化的加持下,整个世界不断修正这些影响。

过去一年,全球都在抗击新冠肺炎疫情。人脸识别等人工智能技术在公共卫生方面发挥了非常大的作用,比如对整个疫情的建模、预估,对疫情的追踪、检测、数据分析,也包括无人送餐、智慧医疗设备等。

"AI+医疗",本就是人工智能未来的大趋势之一。大健康、医疗诊断、新药研发均在人工智能的发展中得到提升。

除了医疗领域,后疫情时代的人工智能还会在哪些方面有

位于两江新区的超声医疗国家工程研究中心,以数字医疗服务全球患者
China's National Engineering Research Center of Ultrasound Medicine in Liangjiang New Area (Chongqing) serves patients worldwide with digital healthcare

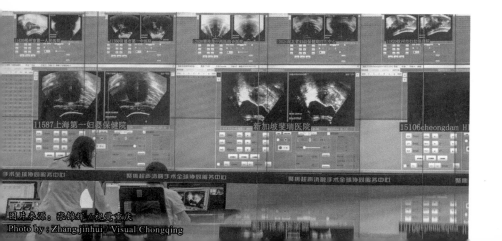

所突破，出席 2020 线上智博会的产业精英们带来了全新的思考。

在百度创始人、董事长兼首席执行官李彦宏的眼中，数字智能技术将从根本上改变我们所熟知的各行各业：医疗健康、工业制造、教育、金融、交通、城市管理等。目前，百度在深度学习、自动驾驶、智能交互、人工智能平台等方面都已有布局。

人工智能技术将为产业智能化服务，而产业智能化的未来，会消灭交通拥堵，提高生产、工作效率，减少资源浪费，实现智能的便民服务，建立更加文明、安全的智能社会，让人振奋而期待。

华为无线网络 5G 产品线总裁彭红华则认为，眼下最火热的 5G 技术，将会催化一切。5G 将连接全场景，实现万物互联，让人工智能无所不及、让计算无处不在、让云触手可及，加速千万行业数字化，创造新价值、新机遇。

高通公司中国区董事长孟樸认为，未来 5G+ 人工智能 + 边缘云，将变革教育方式，加速医疗变革和汽车行业变革，提升工作效率，构建柔性制造系统。同时，还可构建无界 XR(扩展现实) 生态，实现个性化购物体验。[1]

最重要的是，智能化带来的 X 效应，将给众多行业带来巨大变革。

英特尔公司全球副总裁兼中国区总裁杨旭认为，"智能 X 效应"会带来更多的数字增值服务[2]，特别是疫情后，在线办公、在线学习、在线购物等，会从临时性的应用转化成生活的刚需。面对如此庞大的经济结构转型，关键的基础技术支撑将尤为重要，而未来的不确定性，也将有迹可循。

---

[1] 谢艺观，中新网，《中国已建成 5G 基站超 50 万 5G 融合应用加速推进》，2020 年 9 月 16 日。

[2] 腾讯科技，《英特尔中国区总裁杨旭："智能 X 效应"带来更多的数字增值服务》，2020 年 6 月 23 日。

第 5 节　全球疫情中，中国采用人工智能实践抗疫

> 理论所不能解决的那些疑难，实践会给你解决。
> ——路德维希·安德列斯·费尔巴哈

人类社会之所以能持续向前发展，核心驱动力来自人类自身解决问题的能力。遇到问题，主动寻求解决方案，是人类的本能之一。

而当人类拥有了人工智能之后，那些纷繁复杂到让人仿佛束手无策的难题，正在以更快的速度被破解。

2020 年，新冠肺炎疫情成为全球性的黑天鹅事件，而中国率先采用人工智能技术实践抗疫，这个过程给全球带来了诸多宝贵的经验。

无接触服务兴起

从来没有哪个年份，比 2020 年更让人们手足无措。

人们从来没有想过，连接触电梯按钮的方式都能成为施展创意的舞台，抽纸、牙签、保鲜膜、便利贴……都成为人们避开与电梯按键直接接触的神器，但仍没有从根本上解决问题。

于是，百度联合海信地产、碧桂园等合作伙伴推出了语音呼梯的无接触乘梯方案，利用 AI 语音识别实现零接触的电梯

控制，让出行更便捷、安全。此方案目前适用于市面上绝大多数电梯，同时兼容原来的梯控系统。

中国在抗疫初期，采用严格的"封城"政策，控制疫情传播。但生活还要继续，因此催生出一大批无接触服务，包括居家办公、在线教育、线上买菜等。

这场突如其来的疫情给全国各城市的抗疫工作、餐饮休闲、商超便利、文化娱乐、旅游出行等带来了严峻挑战，仅仅依靠人力是远远不够的。尤其是在保证人们日常生活方面，快递企业与外卖企业一早行动起来，推动"无接触配送"，送货机器人、无人物流仓、智能云仓系统等都在探索中投入使用。

在疫情最严重的武汉，京东物流迅速完成机器人配送的地图采集和机器人测试，从各地抽调配送机器人驰援武汉。一直处于研发测试阶段的物流无人机，也提前上岗，为已经封闭的地区提供物流服务。

重庆市商务委则联合美团点评，推出了249家餐饮门店目录，全面推行"无接触点餐"服务，加大餐饮外卖供应，为百姓、

2020线上智博会场馆外，自动驾驶清扫车兼具消毒功能
The automatic sweeper disinfects the area outside the exhibition hall of the 2020 SCE Online

复工复产企业员工提供安心餐食。[1]

针对中小企业复工复产，科技企业运用云计算大力推动企业上云，重点推行远程办公、居家办公、视频会议、网上培训、协同研发和电子商务等在线工作方式。运用"互联网+"技术，保持无接触办公是对抗疫情、保持经济的重要方针策略，也将会带来未来办公方式的整体变革。

线上招聘成了企业引进人才的首选形式，招聘活动全部转到线上，实现就业服务不打烊、网上招聘不停歇，通过在线语音、视频的形式进行面试。

除了人们的衣食住行以外，无接触服务辐射到各个行业。作为复工复产黏合剂的金融行业，也在动荡中重塑。银行业作为金融服务的提供者，在充分融合金融科技的基础上，通过运用人工智能、大数据、云计算等技术，以减少或避免与客户直接接触而开展的新一代金融服务，加速数字化转型，融入新的业务场景，提升"无接触金融"服务能力。

与此同时，强力推动制造企业与信息技术企业合作，深化工业互联网、工业软件、人工智能、VR/AR等新技术应用，推广协同研发、无人生产、远程运营、在线服务等新模式新业态，加快恢复制造业产能。

科技抗疫的高光时刻

2020年2月4日，发布最严禁令的第二天，杭州市余杭区委办开会明确提出要建立一套数字化方案，并且要做到三个全，即"全人群覆盖+全流程掌办+全领域联防"。

2月5日凌晨5点，第一个版本诞生，之后就开始了每半

---

[1] 黄光红，重庆日报，《重庆249家餐饮门店推行"无接触点餐"》，2020年2月18日。

小时一迭代的优化过程。通过手机定位、运营商数据以及疫情散播的大数据比对，可以用红、黄、绿三种颜色的二维码，呈现持码者的防疫健康信息，并提供其他附属功能。

余杭区从杭州疫情的"重灾区"，到杭州复工最快区，转变过程仅一周时间。这要归功于依托人工智能技术开发的这套数字化解决方案。

随后，余杭的试行经验，也就是我们每个人都熟知的"健康码"，在40天内普及全国。不论是出入各个城市，还是进出小区或者医院，健康码都是最有效的通行凭证。

在不同城市，健康码结合当地的防疫情况不断演化。在重庆，腾讯将"渝康码"与"乘车码"合二为一，演化成"健康乘车码"，实现了各种交通工具之间的"一码通行"，一次刷码就能完成健康核验和车费付款，让出行更加便捷。

在重庆洪崖洞景区，志愿者正在帮助游客填写"渝康码"
Volunteers help tourists fill the "Yukang Code" in Hongya Cave (Chongqing)

图片来源：龙帆 / 视觉重庆
Photo by : Long Fan / Visual Chongqing

疫情期间，"健康乘车码"成为3 000万重庆市民必备的"出门条"和"通行证"。

中国之所以能够在全世界范围内最快实现复工复产，公共场所下的测温技术创新发挥了关键作用，尤其是在医院、火车站、机场等人群密集的区域，不但测试结果必须更准，而且测试效率必须更高。在这种情况下，5G技术由于速度快、延迟低的特性，也派上了用场，许多地区的医院和火车站都采用了5G热力成像测温系统。通过红外线体温检测摄像头与5G无线接入设备相结合，在无接触的情况下，能在10米以内，秒速检测体温。

百度AI还基于人脸关键点检测及图像红外温度点阵温度分析算法，实现对一定面积内乘客的额头温度进行检测，即便是佩戴帽子和口罩也能够快速筛查。在地铁、高铁等需要进行大量体温监测的场景中，1分钟内最高可以单通道内检测200人。乘客几乎不用停留，避免了人员拥挤，且温度识别误差仅为±0.3℃。这种非接触式的测温方式，大大降低了交叉感染的风险。

在5G时代，各地的医疗力量不必亲赴一线也能救死扶伤。疫情期间，昆明医科大学第一附属医院启用首个AR/5G的三维数字化远程会诊系统。就算患者远在千里以外，医生只要戴上VR眼镜，一个三维立体的新冠肺炎患者肺部，就会出现在眼前，肺部病灶一目了然。

此前，华为为火神山医院捐赠的远程医疗平台可以直接对接远在北京、上海的优质医疗专家。该方案由"华为TE20视频会议终端+5G-CPE+智慧屏+华为云"组成，系统支持1080P高清画质。

华为云WeLink如今开放了"远程医疗方案"，包括提供远程诊疗、远程探视、远程会议、病案收集、定向推送等五大功能。通过"远程诊疗"和"远程探视",专家医生即可进行全方位的高清远程会诊指导,及时诊断和医治,群众也可以通过视频会议远程探视家人,避免感染风险。

抗新冠肺炎疫情,没有国界线。在全球性的公共卫生危机面前,中国企业在行动,通过人工智能向海外输出检测和诊疗能力,为全球抗疫合作和经济复苏贡献力量。

后疫情时代,人工智能仍是第一主力

人们总是对人工智能充满各种期待,而如何落地应用,一直是人工智能创新项目发展的关键。

在此次新冠肺炎疫情中,人工智能应用基本全程覆盖各个环节,尤其是在疫情监测分析、人员物资管控、后勤保障、药品研发、医疗救治、复工复产这六个方面,大幅节约了人工成本、减少了人力资源消耗,提高了效率,并极大减少了病毒感染传播的风险。

在抗疫初期,人工智能虽然大量投入使用,但稍浮于表面,并没有达到人们对人工智能的真正期望。在社区防控和一线抗疫上,人力还是绝对主力。

而且新冠肺炎疫情爆发突然,数据积累不足,人工智能技术在病毒传播扩散途径检测、病毒源头的追溯等方面表现平平,并没有发挥出应有的作用。

虽然在疫苗的研发过程中,也用了一些大数据和人工智能工具,但起到核心作用的还是一线医疗人员,以及进行制药研究、疫苗研发的科学家。

在人们对人工智能的期望中,"AI+医疗"是一个大方向。但很显然,目前的人工智能在医疗行业还有很大的发展空间,比如对蛋白质结构的预测、对基因机理的研究、对现有高通量测序技术[1]的探索。

未来，人工智能将被许多其他医疗领域加速采用，人工智能技术将被用来处理癌症等大量积压的其他医学问题，而不仅是用来应对病毒传播。

------ ······

陆军军医大学陆军特色医学中心（大坪医院）完成MR全息投影下人工关节置换术
Doctors of Army Medical Center of PLA(Daping Hospital) finish the artificial joint replacement under MR and holographic projection

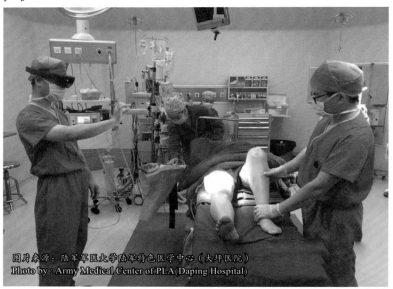

图片来源：陆军军医大学陆军特色医学中心（大坪医院）
Photo by : Army Medical Center of PLA(Daping Hospital)

---

[1] 高通量测序技术：又称"下一代"测序技术，以能一次并行对几十万到几百万条DNA分子进行序列测定和一般读长较短等为标志。

新冠肺炎疫情给我们的生活、工作和社交方式带来了巨大影响。此前，稳定和强劲的数字化趋势虽然已经在社会的许多方面显现，但 2020 年，全球掀起了一场数字化热潮。

后疫情时代，技术正在以惊人的速度发展，不断定义和刷新下一个前沿领域，人工智能仍是改变我们生活、工作和娱乐方式的第一主力军。

在万物互联的未来展望下，更智能的大数据分析、更及时的检测和预防、更具黏性的客户关系、更智能的决策能力，是人工智能最迫切的目标，而认知智能技术和情感计算都将迎来新需求，人机共存已近在眼前。

人工智能仍在成长，离全球普及或许尚早，但不妨碍我们的想象与期待。

第4章

成果：科技智能的前沿创新

不同的场景，需要不同的创新；不同的变革，创造不同的成果。多少领域的分头探索，多少难题的各个击破，一次次的重复实验，一轮轮的跨界连接，终于让人工智能从街巷走出，向广场集中，也终于汇集到一个属于时代的舞台。

智能材料、工业互联、自动驾驶、智慧文旅、智慧防疫、智慧金融、智慧政务……当我们在智博会的舞台上，一一检视来自不同领域的人工智能创新成果，才发现众多人工智能的创新成果，共同撑起了一个全新的时代。

第 1 节 黑科技汇聚，
从智博会眺望时代未来

追上未来，抓住它的本质，把未来转变为现在。

——尼古拉·加夫里诺维奇·车尔尼雪夫斯基

让未来现在就来，是整个世界的期待。

全球每天都在举办各种不同的科技产品发布会，这些新科技像是全世界的科技创新者对未来的集体创作。然而，并不是每一个人都能用这些分散的创新碎片拼接出未来的全貌。

始于 2018 年、连续举办了三届的智博会，汇集全球科技创新，就像是一场以未来为主题的饕餮盛宴，吸引人们前来"品尝"。

每一届智博会上，新技术新理念层出不穷。这些科技推动着社会各领域向智能化加速迈进，更创造了人类与未来世界同步沟通的全新窗口。

2020 线上智博会群英荟萃，吸引了累计 551 家国内外单位线上参展，举办各类论坛共计 41 场。涉及 5G、区块链、工业互联网等 20 个领域。其中，世界 500 强企业占比 8.7%，中国 500 强企业占比 10.3%。

2020 线上智博会数字经济百人会现场
Globalized Digital Economy 100 Forum at the 2020 SCE Online

......

  智能工厂、5G 远程驾驶、智能材料、双腕机器人等竞相亮相智博会，硅基光技术、L4 级自动驾驶中巴车、8K Micro-LED 等一批新技术、新产品也在会上首发，覆盖医疗器械、汽车等领域的十大工业互联网标识解析二级节点正式上线。

  对比前两届，2020 线上智博会的展示成果在应用规模上已有大范围的延展，其中很多创新已经从实验室走向市场，科技早已渗透到人们的日常生活中，不再遥不可及。

  这些来自全球的最新科技，已经成为未来技术发展的风向标。

  新一代人工智能技术的突破和应用，进一步提升了制造业数字化、智能化、网络化的水平，也让创新速度加快，推动新技术应用得更广。

2020线上智博会上，3位诺贝尔奖及图灵奖获得者、49位两院院士、443位知名专家和行业精英围绕人工智能、5G、区块链、工业互联网等前沿热点话题，碰撞思想、交流成果，新思想、新理念、新观点在全球扩散，激荡出关于未来的畅想。

在不久的将来，可以飞行的汽车、脑机互联的接口、空无一人但繁忙生产的工厂，这些在今天看来还颇为科幻的场景，或许将变得稀松平常。未来50年，人类生活将被人工智能重新塑造，而一切的改变也都能在今日今时找到创新的源头。

机器变得聪明，系统更加流畅精准，泛在连接无处不在，智能时代的进程已经不可逆转，机器越来越像人，甚至超过人，而人呢？可以继续发挥想象力，把精力集中在更有意义的未知探索上。

或许，在未来，人人都是程序员，人人都是设计师，人人都是创造者，技术与机器都任由我们奔腾的思绪去改造。

智慧之城图景渐明，礼嘉5G馆引领我们见证奇妙时代到来
Lijia 5G Pavilion leads us to witness the arrival of the wonderful smart city

第 2 节　从智能材料到自动驾驶，制造业正在加速变革

不要害怕新的竞技场。

——埃隆·马斯克

"高智商"的产品，离不开"高智能"的制造。

在新一轮科技革命和产业变革中，智能制造已成为世界各国抢占发展机遇的制高点和主攻方向。

借助传感器、物联网、大数据、云计算等新技术，工业制造行业迎来变革时刻，工厂连上云端，产品有了大脑。

智能材料的研发

过去的机械是冰冷的。如今，贴上"传感器"，机械就有了"神经系统"；贴上"作动器"，机械就有了"肌肉"。曾经冷冰冰的机械仿佛成了有血有肉的"生命体"，在这背后，就是智能材料的作用。

智能材料也叫机敏材料，是一种能感知外部刺激，能够判断并适当处理且本身可执行的新型功能材料。这种材料把高技术传感器或敏感元件与传统结构材料和功能材料结合在一起，

赋予材料崭新的性能，使无生命的材料变得似乎有了"感觉"和"知觉"，并具有自我感知和自我修复的功能。

2019智博会开幕之际，诺奖（重庆）二维材料研究院正式注册成立，由2010年诺贝尔物理学奖获得者、石墨烯发现人之一的科斯提亚·诺沃肖洛夫教授担任名誉院长并牵头组建团队，致力于基础研究、工业项目和标准化等工业服务。

到2020线上智博会召开时，不到一年的时间，研究院已组建起20余人的研究团队。在线上智博会上，科斯提亚·诺沃肖洛夫代表研究院向外界展示了他们在智能材料方面的最新研究成果。

2020线上智博会上，科斯提亚·诺沃肖洛夫教授通过全息投影的方式发表主题演讲

Professor Kostya Novoselov gives a keynote address by holographic projection at the 2020 SCE Online

图片来源：罗斌/视觉重庆
Photo by : Luo Bin / Visual Chongqing

智能材料是继天然材料、合成高分子材料、人工设计材料之后的第四代材料，是现代高技术新材料发展的重要方向之一，将支撑未来高技术的发展，使传统意义下的功能材料和结构材料之间的界线逐渐消失，实现结构功能化、功能多样化。科学家预言，智能材料的研制和大规模应用将导致材料科学发展的重大革命。

短短两年时间，诺奖（重庆）二维材料研究院对二维材料的基础研究已经取得重大突破。研究院对新型的二维材料生产进行研究，不仅实现纯二维晶体生长，还能兼顾合金生长，并尝试将二维材料和其他材料及晶体相结合，例如聚合物和聚合电解质，让这些材料可以应对智能时代更多不同的挑战和任务。

为了应对新冠肺炎疫情，研究院已着手研究智能抗病毒涂层，该种涂层将能通过智能可编程响应来杀死吸附在表面的病毒，对包括新型冠状病毒在内的多种病毒都非常有效。

但这不是此项研究的终点，研究院还将在材料科学领域广泛使用机器学习技术，研究二维材料晶体及其合成物，进一步改进材料，探索创造出一种具有可编程功能的新材料。

目前，研究院在智能纺织品、热管理、电信、光子学、表面等离子体光子学等应用上投入了大量精力，在智能膜和智能涂层方面表现良好。

适应性智能材料能满足多种功能需求，并且能积极地适应外部环境，将在智慧城市、机器人、人工智能、先进的医疗保健、水处理和电信等许多领域得到应用。

"我们希望我们的材料是有生命的，能够像生命系统一样运行。我们相信，我们的适应性智能材料将在智慧城市、机器人、人工智能、先进的医疗保健、水处理和电信等许多领域得到应用。"这是科斯提亚·诺沃肖洛夫的目标。[1]

工业互联网的发展

一段时间以来,工业4.0被广泛认为指引着制造业变革的方向。与此同时,工业互联网也被寄予厚望,工业互联网标识解析体系对工业互联网发展起到重大作用。

在2020线上智博会上,中国信息通信研究院对标识解析国家顶级节点进行了全面展出。早在2018年12月1日,工业互联网标识解析国家顶级节点(重庆)在重庆启动建设,这一国家级工业互联网基础建设有望加快西部地区工业互联网的发展节奏,提升基础服务能力。[2]

简单来讲,工业互联网标识解析,就是为众多联网机器、设备、传感器发放唯一编码的"身份证",就像我们每个人都拥有一张居民身份证一样,有了标识解析系统,就可以对机器、设备、传感器进行唯一性的定位与信息查询,以实现全球供应链系统和企业生产系统的精准对接、产品的全生命周期管理和智能化服务的实施。

而标识解析国家顶级节点,就是一个国家或者地区内部最顶级的标识解析服务节点,能够面向全国范围提供域名、标识、区块链的基础服务和资源管理功能,国家顶级节点既要与各种标识体系的国际根节点保持连通,又要连通国内的各种二级及以下其他标识服务点。

---

[1] 杨野,上游新闻,《诺奖得主科斯提亚·诺沃肖洛夫:用智能抗病毒涂层应对新冠疫情,首次试验取得喜人成效》,2020年9月15日。

[2] 于宏通、何宗渝,新华社,《工业互联网标识解析国家顶级节点(重庆)在渝启动》,2018年12月3日。

2020线上智博会上展示的"国家标识解析运行监测平台"
"National Identity Analysis Monitoring Platform" at the 2020 SCE Online

作为中国最顶级的标识服务节点,国家顶级节点的设立意义重大,它不仅能面向全国范围提供顶级标识解析服务,以及标识备案、标识认证等管理能力,还可以在国际根节点与二级节点之间承上启下,实现连通。可以说,国家顶级节点在整个工业互联网标识解析体系中占据基础设施地位。

截至2021年5月底,中国的工业互联网标识注册量迎来"爆发节点",国家顶级节点共开设了五个,分别位于北京、上海、广州、重庆、武汉五市,已上线二级节点134个,覆盖23个省区市的28个行业,接入企业超过15 000家,全国标识注册总量已突破200亿,日均解析量达到1 200万次。[1] 重庆顶级

---

[1] 黄舍予,人民邮电报,《工业互联网标识注册量超200亿,"加速赋能"效应凸显》,2021年6月1日。

节点于 2018 年 12 月上线，成为重要的数字经济新型基础设施。据中国信通院工业互联网与物联网研究所统计，截至 2021 年 5 月 21 日，国家顶级节点（重庆）标识注册总量达 6.31 亿，累计标识解析总量 3.34 亿次，已接入 19 个二级节点，接入企业节点 1 099 个[1]。

工业互联网将中国的制造业带入一个崭新的时代，正式将"个性化定制"介绍给全中国的生产制造商、消费者，C2M（用户直连制造）模式成为新的市场风口。

阿里云与重庆飞象工业互联网有限公司共建的基于工业互联网的 C2M 产销协同平台

C2M Platform jointly built by Ali Cloud and Feixiang Industrial Internet Company based on Industrial Internet

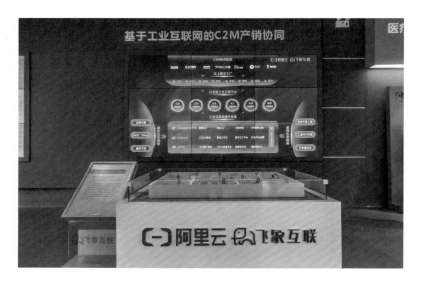

---

[1] 夏元，重庆日报，《国家顶级节点（重庆）标识注册量达到 6.31 亿》，2021 年 5 月 30 日。

在重庆飞象工业互联网有限公司和阿里云共建的 C2M 产销协同模式里，大批量生产与个性化定制实现了数据对接，平台通过淘宝销售系统连接终端客户、经销商、旗舰店等，直接获得订单，实现淘工厂与销售端信息协同。

在 C2M 模式中，从消费者发起个性化需求，到最终商品的交付，中间要经历产品设计、研发、制造、物流、服务等诸多环节，涉及从消费平台到制造工厂的多个场景切换。如果没有工业互联网在其中充分发挥连接与协同作用，显然 C2M 只能是纸上谈兵。

海量的用户订单通过工业互联网平台实时传回工厂，基于 MES 智能决策、自动安排生产，并将信息自动传递至各个工位、工序、生产线及各供应链环节，实现同步生产，并最终完成个性定制产品出厂。

淘工厂快速处理订单，智能生成计划单，同步零部件供应链采购端，根据库存、设备状态、订单排程等，准确下达生产指令，实时监控生产过程。通过智能物流实现智能出库，运输状态实时监控，客户签收后，进行对账结算，形成数据闭环。

而且，下单用户可通过手机、电脑等终端，实时监控产品的生产状态、进度，实现生产制造的可视化。

自动驾驶的成果

自 2010 年谷歌宣布自动驾驶汽车研发计划以来，跟随者与日俱增，苹果、百度、阿里等众多科技公司都争先投入其中。

11 年后的今天，不光是科技公司，新旧车企也纷纷加入，蔚来、理想、小鹏陆续面市，小米、百度相继官宣造车。

在中国，百度早在 2014 年就宣布研发自动驾驶技术，经过 6 年的试错和积累，在 2020 线上智博会对外发布了 L4 级自动驾驶公交车 Robobus。

自动驾驶小巴士
Self-Driving Vehicles

　　自动驾驶公交车一经亮相，便惊艳全场。公交车长5.9米，核载19人，红色漆身的外观，头顶"大眼睛"、两侧"招风耳"自带科技光环。"大眼睛"是自动驾驶小巴士的单目相机，"招风耳"是激光雷达。该车一共配备了4个激光雷达，2个毫米波雷达，7个单目相机。这些设备帮助车辆探测在行驶过程中的各类信息，360度无死角的设计，感应更周全。

　　该车在正常行驶时，遇到横穿的行人或车辆，可以自动识别并停车礼让。此外，自动驾驶公交车还可以轻松识别红绿灯，完成直行、转弯、掉头等操作，能在自动驾驶过程中注意到前后左右的所有情况。为了适应公交场景的特殊性，百度的自动驾驶公交车在站点及吞吐量方面结合实际使用做了额外设计，具备精准泊车能力，能实现精准靠站，轻松应对公交站场景及更为复杂的城市道路路况。除此以外，自动驾驶公交车还有深度学习功能，每天在同一条路上重复行驶，能对道路状况、事故多发点进行分析记录，从而提前预判可能发生的情况。

按照计划，位于永川区的西部首条自动驾驶公交线路于 2020 年 9 月通车运营。重庆人民就此体验到了百度的自动驾驶公交车 Robobus 和自动驾驶出租车 Robotaxi。另外，由重庆永川区人民政府、重庆车检院和百度共同建设的"西部自动驾驶开放测试基地"也正式投入使用。

西部自动驾驶开放测试基地是中国西部地区应用场景最丰富、自动驾驶车辆规模最大的开放测试和示范运营基地，也是中国首个 L4 级自动驾驶开放测试和示范运营基地。基地位于重庆永川中心城区，基于车、路、云、图全面协同的建设模式，全方位部署了 5G 通信路网环境，构建立交、隧道、桥梁等 30 余个山城典型开放道路测试场景，可同时承载 200 台智能驾驶汽车开展测试。

目前，百度已获得 10 张重庆自动驾驶载人测试运营牌照，在基地内开展了 5 种应用场景、20 台 L4 级自动驾驶车辆的测试和示范，包括分别与一汽红旗、福特林肯合作研发的自动驾驶出租车，与厦门金龙合作打造的自动驾驶公交车，以及与庆铃汽车合作打造的无人驾驶环卫车等。

基地同时还全面开放百度 Apollo 自动驾驶测试云控平台，实现了路网环境感知、路侧边缘计算、车路实时信息交互等功能，全面整合重庆车检院国家机动车检测能力平台，在西部率先构建起集"虚拟仿真+封闭实验+开放测试"为一体的自动驾驶全链条检测服务体系，成功获批重庆国家新一代人工智能创新发展试验区十大应用场景。

西部自动驾驶开放测试基地作为西部领先的智能网联创新

样本，加速中国人工智能和智能交通的腾飞，为国内外自动驾驶整车企业、系统解决方案企业和零部件企业，提供良好的研发测试服务和丰富的示范应用场景，共同呼唤智能时代的到来。

智能制造让产品更加智慧

新的时代，不管在精神上还是物质上，人们都有了前所未有的自由。

而自由的注脚由人工智能、5G、物联网等写下，新一代信息技术给人们带来了更迅捷的物流，更高效的服务，更智慧的产品，旧的产品与服务在科技的助推下，以新面目迎接人们惊艳的目光。

在老重庆人眼中，段记西服泛着旧时光的味道，新世界似乎已经没有可以载它的船。但遇到人工智能以后，段记西服重新找回昔日荣光。

重庆段记西服的人工智能服装定制平台，让试穿变得智慧。通过三维扫描仪，扫描和生成特定姿势的人体模型，利用人体重构技术为扫描人体自动加入骨骼，使其能够虚拟着装和展示。而且，加入布料仿真引擎，快速将二维衣片进行网格划分，通过计算机图形库在三维环境下呈现出和真实衣片一样的物理特性，展示出真实衣服穿着在真人身上的效果，包括褶皱、凸起等细节。

在这个定制平台的试穿，和到线下实体店试衣服没有区别。段记西服可以通过人工智能识别出缺陷，快速地修改版型，裁剪出最适合消费者体型的服装，实现服装个性化定制。

从智能材料到工业互联网，再到工业机器人，智能制造给制造业带来一次次腾飞的机会，新一代显示技术 8K Micro LED 也在智能制造的助阵下应运而生。

第 4 章　成果：科技智能的前沿创新

图片来源：龙帆／视觉重庆
Photo by：Long Fan／Visual Chongqing

2020 线上智博会"智慧重镇"展馆内，工作人员正在展示 AI 人工智能服装定制平台

The staff demonstrates the AI customization clothing platform in the Smart Manufacturing City Exhibition Hall of the 2020 SCE Online

2020 线上智博会现场，重庆康佳光电技术研究院有限公司现场展示了 236 英寸 8K Micro LED 大屏，它是全球图像分辨率最高的百英寸 8K Micro LED 显示产品，有超过 9 900 万个区域控光，峰值亮度 3 000nit，对比度 1 000 000∶1，同时搭载全球领先的驱动方案 +5G 通信模块。[1]

基于对重庆大力培育"芯屏器核网"全产业链发展理念的认同，主营业务 ALPD 激光显示技术的峰米科技，选择落户重庆。在 2020 线上智博会现场，峰米科技展示了"峰米 4K 激光影院 Max"，该产品画面亮度高达 4 500ANSI 流明，领先的 HDR10+

---

［1］刘政宁、曾清龙，人民网，《黑科技满满　璧山企业将亮相 2020 线上智博会》，2020 年 9 月 12 日。

及 HLG 解码技术，带来层次更加细腻的画面明暗对比；此外，0.25∶1 的大景深超短焦镜头设计，可以轻松投出最大 200 英寸的巨幕画面，还原更真实的影院级感受，真正让用户实现"把影院搬回家"。

事实上，在普通人的日常生活以外，新一代信息技术在能源领域也有不俗表现。中国船舶重工集团海装风电股份有限公司自主创新研发了面向陆上和海上风场的风电设备 LiGa 大数据平台，可以实现智慧风电远程运维管理，让能源更加智慧。

在 LiGa 大数据平台上，大量的风场运行检修数据、气象数据与风电场管理经验产生化学反应，在风资源评估及宏观选址、大数据采集存储管理、风场智能运维、风电场后评估、数据挖掘分析、故障诊断及预测建模分析等整体解决方案上智慧决策，并实现风电场运行状态的远程实时监视、风机运行生命周期的智能健康管理、产品分析辅助决策等。

5G 技术逐渐走向成熟后，远程服务有了坚实的保障，多家制造公司都在探索 5G 远程运维、5G 智能工厂、5G 远程驾驶等业务。智能制造、智能农业、智能医疗、智能交通等新业态不断涌现。

发展以人工智能技术为核心驱动力，以跨界融合为典型特征的智能产业，促进经济社会转型升级已经成为全球各国共同的战略选择。

而在工业制造领域，这早已成为全球共识。

第 3 节 智博会上的新世界，生活有了"智"的飞跃

科学不问现在和过去，是对一切可能存在事物的观察，预见虽然是渐进的，然而它是对即将发生事物的认识。

——列奥纳多·达·芬奇

未来的智能时代到底会是什么样，一千个人心中有一千种设想。

在 2020 线上智博会上，无论是展馆里展示的黑科技，还是礼嘉智慧公园搭建的智能生活场景，都将即将到来的智慧生活展现得淋漓尽致。

5G 远程驾驶、智慧社区、无人餐厅、未来学院等场景都已触手可及，人们的生活有了"智"的飞跃。

文旅遇上数字化，一台手机走天下

如果说智慧社区的很多创新，是人们日常生活中倍加期待的"意料之中"，那么智慧文旅的诸多实践，可以说是大众意料之外的惊喜。

文旅的体验感不但不易量化，还涉及交通、住宿、游览、

餐饮等方面，怎么兼顾好这些方面从而得出一个最佳方案，是一个挑战，尤其在后疫情时代。

腾讯可谓国内智慧文旅方面的佼佼者。早在 2016 年，腾讯文旅就做出了"互联网＋文化＋旅游"的整体布局，从龙门石窟到一部手机游云南，从数字故宫到数字敦煌，已经积累了上百个创新案例和落地服务。

2020 年 5 月 15 日，中国联通与腾讯在京联合发布了《2020 中国智慧文旅 5G 应用白皮书》，在技术研究与场景展望的基础上，聚焦 5G 文旅行业解决方案和落地实践，全面展示了 5G 促进文旅行业增长、融合、可持续、创新及合作的能力。

后疫情时代的旅游业不是简单地恢复过去的旅游模式，而是要有新思维、新动能和新模式。

在 2020 线上智博会"智慧名城"展馆，腾讯 T-DAY 用智能科技和互动创意，利用重庆特有的两江四岸的元素，结合山水城市特点，打造了一座专属重庆的未来智慧城市，把山水重庆元素和智慧生活场景融为一体，激发公众对智能科技的兴趣，增进对智慧城市的了解。

穿过波光粼粼的展厅入口，耳边响起川江号子，参观者将穿越到老重庆的经典场景，感受城市的历史智慧和人文风光。人们只需戴上 VR 眼镜，对"腾讯小微"说出目的地，李子坝穿楼轻轨、武隆天坑地缝、长江索道等网红景点就会随之而来，给人身临其境的体验。

通过时光隧道，从过去到未来，大数据中心、5G 网络、云服务资源等构造起一张城市运行的数字大网，数据川流不息，科技遍布城市的每一个角落。

腾讯以广泛的国内社交用户群为基础，结合自身业务发展过程中积累的云计算、大数据、人工智能、安全等基础能力，在全域旅游、景区、文博、特色小镇、主题公园、绿地湿地、

人工智能给山水之城插上智慧的翅膀,重庆美景尽收眼底
Artificial intelligence accelerates the development of Chongqing and makes the land of natural beauty at your fingertips

会展等细分方向形成专业的数字化服务能力,并推动将5G等信息技术应用渗透到各个智慧场景中。

除了腾讯以外,重庆本土的众多智慧文旅"黑科技"产品也在2020线上智博会上大放异彩。银联商务、重庆旅游云、重庆旅次方等多家企业,携带诸多新兴文旅技术应用首次亮相,为广大观众开启从"云"到"端"的丰富体验。

由大足石刻研究院研发推出的大足石刻世界文化遗产监测预警系统,为世界石质文物保护提供了宝贵的"中国经验"。从揭示大熊猫前世今生的重庆自然博物馆古熊猫修复项目到重庆中国三峡博物馆三维古文物修复,再到根据"松石间意"千年古琴开发的"高山流水"古琴科普微信小程序,数智文旅新融合正逐渐走进人们的生活。

后疫情时代，科技为生活添彩

后疫情时代，体温测量已经成为进入公众场所的必要环节，特别是商场、医院、机场、地铁站等人群密集场所。

中国在短时间内，通过人工智能技术，迅速建立起一整套综合应用了极速测温、健康码、行程码等的防范措施，在前面的章节中已有介绍，在这里不再赘述。

对机场而言，防疫工作的难度更高，因为乘客跨区域旅行的范围更广。特别是拥有国际航线的机场，更是要通过多种措施防患于未然，除了要严格做好行程溯源、核酸检测、定点隔离等措施，也要在机场场景内尽可能地减少人们的接触机会。

重庆江北国际机场正在加速推进"智慧机场2.0"建设。随着行李自助托运、无纸化乘机等智能服务的陆续开展，让旅客乘机出行更便捷，也避免了行李托运、换登机牌等环节的集中排队。在T3A航站楼3J岛南侧，16台自助托运行李设备一字排列，旅客只需要扫描身份证等有效证件，便可以在短短一分钟时间内，自助完成登机牌打印和行李托运。行李托运之后，不再需要人工分拣，大数据采集分析技术的使用，将托运行李精准分拣到对应的航班。而且，乘客还可以通过手机实时查看自己的行李信息。

作为目前国内最为先进的分拣系统之一，重庆江北国际机场T3A航站楼的行李系统线体总体长度达到了17公里，整个系统的主要设备总量超过3 000台。首次在国内采用开环式RFID识别应用技术，大大提高了行李分拣的识别率。这套系统2017年9月投入使用后，在国庆节期间就处理了14万件行李，最终成绩是分毫不差。[1]

---

[1] 李相博，新华网，《打探重庆T3A行李分拣系统"黑科技"》，2017年10月8日。

第 4 章 成果：科技智能的前沿创新

国内首个自助行李分拣系统在重庆江北国际机场 T3A 航站楼投用

The first domestic Self-service Baggage Handling System is put into use in Terminal 3 of Chongqing Jiangbei International Airport

后疫情时代，远程办公、在线教育等将继续被保留，而人工智能在其中将继续产生更深远的影响。依靠人工智能、云计算等技术构建创新型运营模式，促进降本增效、业务增长，将成为未来各行业需要重点考虑的问题。

尤其是在医疗板块，科技正在加速促进行业智能化。在重庆山外山血液净化技术股份有限公司的医疗器械行业标识解析应用场景中，一名"患者"身上插满导管，模拟的血液、血透浓缩液和废液通过导管流动。旁边的大屏幕上实时显示着血液浓度等多个临床数据，通过挖掘这些数据，能够实现对每位患者的个性化透析治疗管理。

在 2020 线上智博会，重庆交通大学信息科学与工程学院蓝章礼教授及其团队带来了一款健康检测智能马桶。它可以检测分析尿液、尿流率、体脂、血压、心率等,并且可以结合手机、平板电脑等形成采集端、展示端、云端的平台体系。

眼下,用技术、数据、数据科学等手段推动看病就诊、病历诊断、医疗报销等就医流程的数字化、智能化,成为医疗信息化发展的重要发展方向。从之前排队挂号、取号、看病苦等数小时,到现在的网上预约、自动分诊、电子病历建设、医保实时结算,看病效率提高,资源浪费减少,医生和病人的满意度都有所上升。

随着人工智能、5G、物联网等技术的发展,医疗与技术的融合也将更加深入,科技为智慧医疗提供了更广阔的想象空间。

礼嘉智慧公园带来新启示

作为智博会永久会址,重庆正在全力打造智慧之城。

而重庆两江新区礼嘉智慧公园作为国内首个智慧公园,展示智能生活的集中载体,人气颇高。园内游客络绎不绝,俨然已成为市民和游客争相"打卡"、贴近自然、养生游憩的体验休闲场所。数据显示,从2019年8月开园,一年时间接待游客就超过50万人次。[1]

公园规划打造陵江次元、云尚花林、极客社区、湖畔智芯、创新中心五大区域,形成"一园五区"的功能结构布局,搭建起了永不落幕的智慧舞台。全世界顶尖的技术、智能的应用争相落户于此,让人们提前"一站式"体验未来城市的智慧生活。

从走进公园大门的那一刻起,智慧生活就已经开始。入口处便有"人脸识别智能储存柜",你只需要通过人脸识别,便

---

[1] 钟旖、韩潇,中新网,《重庆两江新区礼嘉智慧公园:科技云集"智慧"生活触手可及》,2020年10月22日。

能储存物品。入口右侧有一块"脸部分析屏",你只要站在屏幕前,便会显示你的年龄、脸部特征等信息。逛完公园后,可通过智能屏幕了解自己走了多少步、心率情况、消耗了多少能量等。

这,仅仅是智慧体验的开始。

接下来,一条充满"黑科技"的智慧凉道成为重庆炎夏的"救星"。智慧凉道两侧设有智慧喷雾降温系统,只要进入喷雾区,智能系统将识别每个人的体温,根据体温的高低喷洒水雾,不仅可以保持身体的舒适,更营造出"日照香炉生紫烟"的氛围。

────────────────── ......

充满"黑科技"的智慧凉道,可自动降温
The Cooling Path equipped whith black technologies can automatically cool the path

如果你不想走路，园内的无人驾驶车将会成为你的贴心代步工具，车辆可以按照既定路线自动行驶，可以对前方20米以内的物品进行扫描，在距离障碍物5米时，便会做出判断，进行刹车、减速等，自动规避前方障碍物。

如果你饿了，无人小面机、无人包子机、无人碗碗菜机在智慧商业区等你光顾。你只需要扫码下单，通过智能化机械设备煮面、打作料，几分钟之后一道道重庆特色小吃就被送到你的餐桌上。

5G馆外，机器人抬起手臂用钢琴弹奏《我和我的祖国》，优美乐声流淌而出。无人驾驶的清扫车，四处游走保证园区清洁，遇见人还会"主动"让行。你可以通过VR游戏打一场冰球、踢一场足球，甚至骑着5G自行车在大屏幕上欣赏城市四时美景。

艺趣馆、5G馆、云尚体验中心、云顶集市等场馆，云尚花海、智慧步道、云尚观景台等场景，无人驾驶车、机器人钢琴弹奏、VR沉浸式体验设备等体验项目，正在全方位展示未来生活的可能性。

目前，礼嘉智慧公园已在绿水青山间以50个场景立体化展现智慧城市、150种体验全方位探索未来生活，遵循江山湖底色，运用智能化先进要素，汲取智博会优秀成果，构建绿色智能的互动体验场景，打造开放共享的城市活力空间，携超前理念、用世界眼光，打造"智造重镇""智慧名城"展示窗口。

未来，随着重庆城市发展和科技迭代，礼嘉智慧公园还会应用更多智慧场景，让人们的生活实现"智"的飞跃。

理想照进现实，智能时代来敲门

人们在相互祝福的时候，经常会用到"心想事成"一词，只需想一想，就能达成所愿，或许是全人类共通的理想。

而在智能产业时代，人工智能正在将各种"心想"变成"事成"。

毫无疑问，未来已来。理想与现实之间，已无天堑，或许只隔着一条街的距离。

眼下的世界，物联网虽然还未完全连接世间的一切，但已开始局部试验性应用并反馈良好，万物互联只是时间问题。

卫星、无人机、自动驾驶和增强现实这四大创新技术正通过传感器将全球所有设备连接起来，从而创造一个真正的万物互联的社会。

在过去十年中，每个连接设备的传感器数量呈指数级增长，手机上的传感器每四年翻一番。据 IDC 预测，到 2024 年，全球物联网联接量接近 650 亿，是手机联接量的 11.4 倍。[1]

从邮电时代到电信时代，从互联网时代到移动互联网时代，进化的不仅是连接介质，更是连接思维。

当我们早上醒来，只需要戴上增强现实隐形眼镜，它就会记录一天内的每一次谈话，每个人过马路的场景以及我们看到的一切画面。从这个不断观察的数据流中，我们可以使用收集到的个人社交与偏好数据来训练私人 AI。

而所有的传感器，最终都会通过智能互联，实现人与物、物与物的连接和交互。

物联网所产生和使用的数据，囊括了人类社会衣食住行等日常生活的不同方面，同时，也关系到公共管理、生产制造、医疗卫生、交通科技等不同领域。

在未来，每个人都将用上脑机接口，世界已知的信息都在人机交互中完成自主学习，人们从众多的信息中找到喜欢的领域，去探索领域内的未知。那时，人类就可以通过大脑间的直

---

[1] IDC，《中国物联网连接规模预测（2021—2025）》。

接交流交换思想，用思想控制机器更是不在话下，而类脑智能帮助人们处理数据、关联分析，可解决通用场景问题，最终实现强人工智能和通用智能的构想。

有研究者预测，在未来的二十到三十年内，可能会出现能够通过新的图灵测试的、具有通用人工智能的类脑人工智能。类脑智能的成熟可以帮助科学家制造人造大脑，它的存储密度将赶上甚至超过生物大脑，能耗却更低，可以催生更智能的机器人、自动驾驶汽车、医疗诊断等人工智能交互系统。

总之，在物联网无处不在和传感器数以万亿计的现实中，智能时代已经如期而来。身处这个时代的我们，都能看到生活正在产生"智"的飞跃。即便我们不是人工智能的创造者，也应该是智能时代的参与者，迎接一个由物联网与传感器组成的新世界，与我们每一个人都息息相关。

第 4 节　智慧金融变革中，科技带来"涡轮增压"效应

科学的唯一目的，在于减轻人类生存的艰辛。

——贝尔托·布莱希特

说起金融数字化，大多数人脑海里唯一能想到的只有数字货币。但市场的背后，金融科技已经在近几年大范围铺展开来，走进人们的日常生活之中。

2000 年以来，互联网技术的迅猛发展改变了很多行业，古老的金融业也不例外。因为金融行业的数据最充裕，质量最好，这里的"土壤"最肥沃，空间最大，信息技术可以酣畅淋漓地排兵布阵，辗转腾挪。

于是，科技与金融紧密结合，进化出一个新物种。

无接触金融服务

从接触到不接触，人们以为要两年才能实现，金融机构只用了两个月。

新冠肺炎疫情对全球金融体系带来显著冲击，促使全球金融体系在动荡中重塑，行业打造"无接触金融"的共识，让金

融科技大步迈进。两会期间，"无接触金融"再次提档升级，成为新基建的一部分。

疫情期间，人们有贷款、理财、保险索赔等业务需求，却根本没法出门。针对这一痛点，不少银行、持牌消费金融公司等传统金融机构纷纷加速向"无接触金融"靠拢，技术投入、业务转型进行得如火如荼。

中国联通和招商银行共同成立的招联金融，在诞生之时就选择了一条对科技能力要求极高的纯线上发展道路，通过拥抱新技术，率先创新打造了纯线上、轻运营的互联网消费金融经营模式，成为同业内首家全系统去 IOE、首家系统整体上云的公司。

疫情期间，招联金融大放异彩。这一创新模式完全依托互联网获客、经营，没有直销人员，没有客户经理，业务系统整体上云、核心系统去 IOE、公司机房跨省搬迁三件事同步实施并成功完成。

疫情期间，招联金融出动了约 5 000 个智能机器人。[1] 这不仅解决了疫情防控人工需求的压力，也能保障正常经营不受影响。招联金融自主研发的智能机器人具有低成本、多场景、高产出、高效能、易追踪等特点，能以高达 99% 的准确率识别 200 余种用户意图，承担了公司 95% 的客户服务与贷后资产管理工作，有效提升了服务质量和客户体验。

除了招联金融以外，其他金融机构大多也开通了移动端业务，组成手机 APP、微信公众号、社群自媒体、渠道推介等矩阵，实现 360 度全面打通服务渠道。

特别是个人贷款业务，客户"一键触达"，将证件信息、

---

[1] 张玫，人民网，《金融战"疫"服务再升级　招联金融启用约 5 000 个 AI 机器人》，2020 年 2 月 18 日。

资金交易流水等快速提交给平台，平台自动识别信息，完成信息快速登记。平台在线开启智能风控模型，快速评定审核，及时向客户推送签约信息，完成线上签约。批复成功的业务，在当天或次日就能收到款项。

一套流程下来，仅仅一两个小时，解了很多人的燃眉之急。

用科技的力量为中小企业助力

新冠肺炎疫情的爆发对金融业提供快速、精准、"非接触式"金融服务的能力提出挑战，也为金融科技发挥线上化、智能化等优势提供了契机。

疫情对中小企业造成巨大冲击，特别是给融资和风控带来重压，无接触服务金融成为中小企业的救命良药。

对规模小、授信额度低的长尾小微企业来说，它们的资金需求存在短、小、频、急的特点。同时，企业本身又存在高淘汰率、财务信息不规范、缺乏有效资产抵押物等因素，以往就很难在传统授信模式下得到金融支持。疫情的发生，给银行线下网点和信贷人员的工作增加了难度，小微企业的处境更加艰难。

为了解决中小企业融资难融资贵的问题，金融壹账通率先在业内推出"中小企业服务一体化平台—抗疫版"，通过大数据模型协助政府进行宏观经济和中观产业的深度分析，预判行业发展前景，形成企业画像，筛选疫情企业白名单，为精准信贷政策提供依据。依靠智能规划、智能服务、智能产品和智能运营四大"武器"，建设中小企业服务平台和抗疫专区，为中小企业复工复产助力。平台还通过 Askbob 智能政策搜索推荐引擎，实现融资与补贴政策一键匹配申请，已集成 6 000 多条政策，通过 AI 智能语义技术，可以迅速为企业匹配相关政策。

借助云计算、大数据等新技术，金融科技平台得以实现线上客户服务。在疫情引发的线下向线上活动迁移的过程中,金融科技平台可以简化传统流程,减少中介环节,无接触、零距离触达长尾客户,降低信息不对称带来的阻力,为客户提供更高效便捷的服务。

总体来看,在多方共同努力下,2020年小微企业融资取得了"量增、价降、面扩"的显著效果。截至2020年年底,普惠小微贷款余额15.1万亿元,同比增长30.3%,全年共支持小微经营主体3 228万户,同比增长19.4%[1]。

金融科技这个新物种推动了金融业的诸多变革,优化金融供给侧结构性改革,助力银行业数字化转型,赋能区域性银行风控转型与发展,建立开放与共享的新金融体系,帮助金融机构实现"三升两降",即收入、效率、服务质量提升,风险、成本下降。

中国金融科技发展已居世界前列

未来,到底是金融主宰科技还是科技主宰金融?

就中国的情况来看,早几年科技的风头明显比金融更强劲,BAT等科技巨头争先进入金融领域,大量P2P、众筹、网贷公司诞生,给传统金融机构带来了巨大压力。这两年,传统金融机构逐渐缓过神来,开始自己大力推进科技与金融的结合,纷纷设立科技公司,重视科技对金融的促进作用。

汽车的发动机通过涡轮增压,可以提高进气量,从而提高发动机的功率和扭矩,让车子更有劲,也更省油。金融机构通过科技的"涡轮增压",可以提升效率,降低成本,提供更好的用户体验,让金融更美好。

---

[1] 姚均芳,新华社,《余额突破15万亿元!小微企业贷款迅猛增长靠什么?》,2021年1月22日。

对传统银行来说，借助人工智能技术，可以实现智能客服、智能身份认证、智能化运维、智能投顾、智能理赔、反欺诈与智能风控等。大数据技术的应用，可以使银行更广泛地收集各种渠道信息从而进行分析应用与风险管理、精准营销与场景获客。移动互联网技术的应用，大大扩展了银行的市场空间，让人们随时随地进行金融交易。

显然，金融与科技既相互交融，但又各安其位、扬其所长，是目前各方看法的最大公约数。对大多数传统金融机构而言，新冠肺炎疫情催化下的转型，是一次必须面对的"自我革命"。"无接触金融"不只是将固有业务转至线上，本质上意味着经营理念、业务模式、人员结构等方面的联动变革。

2020年12月17日，中关村互联网金融研究院发布的《中国金融科技与数字普惠金融发展报告（2020）》显示，中国金融科技产业发展位居世界前列，2019年金融科技营收规模约为1.4万亿元，同年，中国金融科技融资额占全球比重为52.7%[1]。

总体来看，中国金融科技创新主要以大数据、人工智能和区块链为核心。未来的金融科技，则有可能向区块链偏移。金融科技创新的主力是工商银行、中国银行、建设银行、微众银行等大型机构。

发展无尽头，创新无止境。金融作为市场流通和社会运行的黏合剂，大范围地与新一代信息技术融合后，将反哺科技行业，助力新技术的研发和业务创新，让智能时代更早到来。

---

[1] 潘福达，北京日报，《中国金融科技发展位居世界前列》，2020年12月18日。

第 5 节　生活的便利，从智能政务开始

> 生活在世界上，就有使它更美好的义务。
>
> ——何塞·马蒂

从人们对智能时代有所期待开始，智能政务就成了要预先推动的节点。

热线、大厅、网上办事所等智能政务必然走向更为重要的位置，这不仅仅是直面民生最关键的岗位，同时也是中国政务数据智能发展未来的重要路线。

智能政务与智慧城市之间，存在着直接的因果关系。

街道、路灯、楼栋、天气、民生政策，都在智能政务的管理范围以内。

智能政务的重庆答卷

自 2013 年重庆被列为国家智慧城市试点城市以来，智能家居、物联网、自动驾驶、5G 等技术在重庆生根发芽，智博会长期落户于此，重庆在智能化的道路上一路飞驰。

在智能政务上，重庆一早就开始布局。

自 2014 年 11 月浪潮集团在重庆两江新区建立的云计算中心启动，从启动到运营至今已有七年。在这期间，重庆政务云从硬件到软件再到服务，交出了自己的答卷。

截至 2021 年 1 月，全市"云长"单位达到 110 个，累计推动 2 458 个信息系统上云，上云率由实施前的 26.6% 上升至 98.9%，并通过"数字重庆"云平台形成电子政务云"一云承载"服务体系。[1]

在重庆市江北区智慧城市运营管理中心巨大的显示屏上，数据实时滚动，监控实时在线，全区城市管理系统运行状态一目了然。

智慧城管系统可以全天候、全覆盖地对城市设施、环境卫生等进行监督管理，系统连接有 2.5 万个视频监控点位，覆盖全区所有重点路段。

如果下雨产生路面积水，视频抓拍系统会立刻捕捉到并提示值守人员，值守人员审核信息后通知有关部门前往处理，只需要十几分钟，路面即可恢复正常。

井盖倾斜，传感器自动预警，不用担心路人跌落；路灯不亮，传感器通过电流变化确定位置，维修人员立马赶到……这些在人工排查中难免疏漏的小问题，在系统中无所遁形，处置更加全面高效。截至 2020 年 7 月底，江北区共办理城市管理案件 249 万件，群众满意率达 92.33%。[2]

---

[1] 夏元，重庆日报，《重庆"智慧名城"建设提速　汇聚数字经济企业 1.85 万家》，2021 年 1 月 21 日。

[2] 崔佳、刘新吾，人民日报，《智能化让城市更美好》，2020 年 8 月 29 日。

重庆智能政务,渝快办 24 小时在线
Online Government Service Platform, "Yukuaiban" offers 24-hour government services

不只是江北区,在高新区政务服务中心办事中心,引进的浪潮集团的"智慧政务系统+智能政务自助终端",免去群众来回奔波,线上"一网",线下"一窗",一次性跑完所办业务,群众办事更加便捷高效。

与此同时,高新区还上线了"科学政务",开设了"办事指南"专题,该功能可为企业、群众提供线上查询办事指南和在线申报服务,打造线上"一网"服务,实现统一入口、统一导航、统一认证、统一申报、统一查询、统一互动、统一评价。

另外,浪潮集团还协助开发了"科学政务"微信办事大厅——"口袋版办事服务大厅"。通过微信就能实现排队预约、在线申报、咨询反馈、投诉评价等功能,让政务服务方式更加多元化,让移动政务更加方便群众。

2020 年 12 月，重庆市协同办公云平台"渝快政"上线运行。"渝快政"是全市统筹统建的一体化、智能化、数字化政务工作平台，具备全市政务通讯录、融合通讯、音视频会议、全市公文交换等基础政务协同功能，以及会议管理、会议协同、公文流转、督查督办、事务管理等 40 多项通用办公功能。可确保现在全市党政机关、村社区组织、企事业单位实现"跨层级、跨部门、跨地域、跨业务、跨系统"协同工作，提高政务服务效率。

依托"互联网+"、云计算和大数据，重庆市所有区县已全部完成数字城市管理平台建设。智能化让城市更美好，已成为越来越多山城市民的切身感受。

新加坡电子政务带来的启示

自 1980 年开始，新加坡先后提出了"国家计算机计划""国家信息技术计划""信息技术 IT2000—智慧岛计划""信息通信 21 世纪计划""智能国 2015 计划""智慧国 2025 计划"等规划，在智能政务领域一早就开始了探索。

和重庆以政务云平台为中心接入各区域的政务系统不同，新加坡以 SingPass 账号平台为核心，公民使用 SingPass 账号可以通行所有政府相关的系统，包括 CPF 公积金系统、IRAS 税务系统、ICA 移民厅、MOE 教育部、LAT 交通运输部、房产管理、劳工部等，这几乎涵盖了一个新加坡公民所需的大部分政务服务。

新加坡政务网站几乎包揽了居民日常生活中各项所需，为居民提供"从出生到死亡的全部服务"，这也为新加坡的发展起到了推动作用。

作为建设电子政务的佼佼者，自 2020 年 6 月，部分新加坡政府部门陆续开始安装摄像头。民众无需身份证，只需刷脸就能享受政府部门提供的服务。这些摄像头可以与 2018 年推出的 SingPass Mobile 实现互操作，民众可以利用 SingPass Mobile 在政府的生物信息数据库中注册自己的指纹和脸部信息，之后通过摄像头扫描面部进行比对。

此外，新加坡还有 CorPass 企业管理平台，企业可以在平台上一站式完成政府服务，包括申请补贴、发布招聘、纳税、办理公积金、变更公司资料、申请和取消工作准证等，类似于工商局、社保局、税务局等的合体。

2019 年，新加坡政务部门打造实时 VR 培训平台 HTS2 系统，这是世界上首个结合了实时和虚拟元素的培训平台，该系统既具备沉浸式的特性，同时受训队员还能在物理环境下使用自己的工作站，然后在高度逼真的环境中模拟现实生活的场景，连地标建筑的还原都栩栩如生。

目前，HTS2 系统已经在新加坡警察部队、新加坡民防部队、移民和检查站管理局、新加坡监狱服务处 4 个政务部门展开应用，超过 1000 名警官参与受训。

而新加坡电子政务给智能政务领域带来的最大启示在于包容与开放，数据的透明和数据互联让政务变得更简单和便捷。

智能政务是未来的必然

在新一代信息技术的引导下，政府服务有了全新的特征，包括个性化、云端化、数据化、智能化等。政务运行和政务服务更有针对性和效率，更加理解大众需求，从而开启了一个全新的智慧政务时代。与之前的政府服务相比，智慧政务强调社会服务，更加关注和理解大众个体的需要，更加聪明，更具有预见性。

政府拥有普通企业和个人所无法拥有的大数据资源，而且具有行政力量来获取各种必要的外部社会数据。从政府自身拥有的数据来看，包括工商、税务、公安、交通、医疗、卫生、就业、社保、地理、文化、教育、科技、环境、金融、统计、气象等。从外部数据来看，政府也有能力获得社会数据，如世界经济运行数据、舆情数据、企业报告数据等。政务大数据应用改变了传统政府因不在一线而导致的后知后觉问题，可以通过大数据建模，对社会和经济问题提前预测和预警，实现更有效的政府管理和社会服务。

在新一轮科技革命和产业变革的背景之下，政务云、大数据与人工智能三位一体已成为智能政务的主要方向。比如政务云是目前新技术与政府服务结合最成熟的领域，包括中央政府和地方政府都有大量的应用实例。又如疫情期间开始使用的健康码，现在已成为中国人的出门必备。

智慧政务已有了基础，未来无限可能，我们正在迎来一个全新的政府服务时代。

在重庆市两江新区（自贸试验区）政务大厅，机器人在为市民办理事务

The robot serves citizens in the administrative service hall of Chongqing Liangjiang New Area (Pilot Free Trade Zone)

图片来源：张锦辉/视觉重庆
Photo by: Zhang jinhui / Visual Chongqing

第 5 章

重庆：智能时代的关键见证

　　一座城市，永远不会缺少自己的文化和韵味，也不会缺少自己的商业与生机，前者在城的屹立中沉淀而成，后者在人的流动中随风而来。然而，城市与城市的命运，往往又大不相同：一座普通城市的生命力，主要是看地上有多少热闹的场景；一座伟大城市的生命力，主要是看天边有多么绚丽的蓝图。

　　重庆，在全球范围内，或许不能执智能产业之牛耳，然而近年来的发展总是让人振奋与惊叹。用想象去构筑战略，以实践去触碰未来，重庆城拥抱智能产业之坚定与果决，足以成为众多城市的榜样。

　　重庆，是一座观察全球智能产业发展的瞭望塔，也是一位借助时代机会重新定义自我的领航员。

第 1 节　想象之城：
重庆投身智能产业新浪潮

现实的世界是有限度的，想象的世界是无涯际的。

——让·雅克·卢梭

在我们生活的这个蓝色星球上，人类已鉴定出的生物物种有 170 多万个，实际数目比这要高得多。可是，为什么人类可以超越一切物种，在地球上建立璀璨的文明？

关于这个问题，众多的历史学家从不同的视角给出了完全不同的回答，而新锐历史学家尤瓦尔·赫拉利在《人类简史：从动物到上帝》一书中，给出了一个全新的答案：因为人类拥有创造及相信虚构事物和故事的能力，也就是人类的想象能力。

神明、国家、民族、企业、金钱……所有已经成为基本共识的现代文明构成元素，源头都是人类的共同想象。

人类群聚而居，于是就有了城市，城市汇集的市民想象力的总和，就成为这座城市的想象力。当然，不同的城市，有着完全不同的想象能力。

重庆虽然地处西部，智能产业基础环境并不占优，但先是被列为国家智慧城市试点城市，之后又成为智博会永久举办地，这座城市被赋予了智能时代新的活力。

重庆城，正在努力让自己成为一座智能产业时代的想象之城。

重庆物联网百花齐放

"重庆成为国家智慧城市试点城市！"这是2013年重庆取得的一张新名片。

智慧城市到底是什么？回顾当年的新闻报道，人们都是凭借自己的想象来构建与定义自己心目中的智慧重庆：有的企业开发了手机客户端，将城市的各种信息搬上去，并分为精彩城市、生态城市、特色行业、政府服务、投资城市、名优企业六大板块，认为这就是智慧城市；有的人开发了一套虚拟试衣"魔镜"，认为这也是智慧城市……

对一座城市而言，八年不过是弹指一挥间，时过境迁，重庆在智能产业的布局，早已超越了当时的想象。

早在2010年12月，工信部就已经正式批准重庆市南岸区为"国家新型工业化产业示范基地电子信息（物联网）"，该示范基地也是我国首批国家物联网产业示范基地。得益于重庆成为国家智慧城市试点城市，国内外一批物联网企业纷纷落户南岸区物联网产业示范基地，其物联网产业集聚效应逐步显现。

目前，该基地已形成以龙头企业为基础，大规模物联网运营平台为支撑，产业链大中小微企业协同发展的格局，累计入驻省级以上企业技术中心或研发中心21个[1]，集物联网研发、检测、运营和服务平台于一体，物联网平台种类齐全。

该基地通过开放共享、产品服务等发展方式，在交通、环

---

[1] 饶金兰，南岸网，《南岸物联网示范基地连续两年获评国家五星级称号》，2020年5月8日。

保等行业领域，形成了中移物联网、OneNET 物联网开放云平台等一批大规模运营平台和公共服务平台。目前，该基地以车联网、工业物联网建设为产业内容，正逐步形成集物联网芯片、模组、终端、平台、应用为一体的完整产业体系，实现物联网产业链技术融合创新。

2014 年年底，南岸茶园还到处都是建筑工地，如今引以为傲的智能家居彼时才初具雏形。但那时，聚集在物联网产业示范基地的十几家智能家居公司已经开始头脑风暴，数据交互难以落地，那电力接驳是不是可行？智慧生态太远，是不是可以先研发操作系统？

随着重庆公有云和政务云的建设，智能家居构想中的数据交互得以实现，相应地，智能家居也从前期单纯的远程控制走向半智能的全屋控制，再融入 3D 传感器、物联网等技术，成为今天的智能家居。

在攻关智能家居核心技术的同时，重庆也开始布局智能家居制造的产业链。

在惠达智能家居（重庆）有限公司生产园区内，一条工业 4.0 智能化卫浴生产线已投入运营，产品生产全过程采用自动化物流输送设备，全过程在线生产，各工序间设置自动化立体存储库，产品品种通过 RFID 条码管理系统自动识别，自动按订单出库，生产更加柔性化。

而关于未来智能家居的想象，在建的重庆珞璜智能家居小镇正在挥洒灵感，未来，该小镇将成为智能家居标准制定者、潮流发布者。重庆物联网发展到现在，已成为全国物联网产业重点区域之一，从实时感知长江、嘉陵江水质变化到智慧监控交通状况，从大幅提升企业生产效率到推动产业链创新重构，越来越多的物联网应用在重庆生根发芽、开花结果。

在2020物联重庆融合发展论坛暨会员大会上，2020年重庆市物联网十大应用案例出炉，在智慧交通、智慧医疗、智能安防等领域涌现了一大批典型示范，均已在重庆市试验成功，具备可复制的商业和运营模式，有较强的推广价值。[1]

产业集聚效应下的进化

目前，重庆已经集齐了物联网、人工智能、大数据、5G四张智能时代的王牌，智能时代的大门已在徐徐敞开。

而关于智能时代的想象空间，在四张王牌的加持下，仍在持续变大。

在物联网方面，重庆不但形成了完整的产业链，还有十大成熟的落地应用成功案例；在人工智能方面，重庆建设了国家新一代人工智能创新发展试验区，目前已突破一批关键核心技术，并取得了阶段性成效。如本土企业重庆新安碧捷物联科技有限公司打造的安欣舒智能化安全输液控制系统，包括滴速式输液控制器、护士站管理中心、智能穿戴设备以及ZigBee无线网络四个模块。护理人员身上佩戴的智能穿戴设备可以将所辖病人的液异情况进行及时推送，不论护理人员身在何处，都能及时知晓病人的输液情况。

依托物联网、人工智能、大数据、5G的成果，重庆实施"战略性新兴产业集群发展"工程，迅速做大新兴产业规模。聚焦新一轮科技革命和产业变革战略趋向，加快培育建设一批战略性新兴产业集群，推动新一代信息技术、新能源及智能网联汽车、高端装备、新材料、生物医药、节能环保、软件信息服务等领域集群化、融合化、生态化发展，全力构筑全市经济发展

---

[1] 韩梦霖，新华网，《重庆发布2020年物联网十大应用案例涉多个领域》，2020年12月25日。

图片来源:谢智强/视觉重庆
Photo by: Xie Zhiqiang / Visual Chongqing

工作人员正在重庆永川区兴龙大道实地测试自动驾驶车辆
Workers test autonomous vehicles on Xinglong Street of Yongchuan District, Chongqing

新优势。

"十四五"时期,重庆工业发展目标为工业总量突破 3 万亿元,工业增加值年均增长 6%,占 GDP 比重达到 30%,建成一批千百亿级产业集群。[1]

产业集聚效应下,"产业融合"成为智能产业内的高频词汇,尤其是智能制造的深度实施,推动新一代信息技术与制造业深度融合,加大企业设备更新和技术改造力度,促进产业向价值链高端跃升,让重庆两化融合发展总体水平居全国前列,对外释放智能产业的魅力。

---

[1] 郑三波,重庆晨报,《总量突破 3 万亿,"十四五"重庆工业发展主要任务敲定》,2021 年 2 月 19 日。

与此同时，自动驾驶、智慧医疗、智能政务等领域都在产业集聚效应下，飞速更新换代。目前，自动驾驶已融入5G，衍生出5G远程驾驶系统和自动驾驶汽车两个大方向。智慧医疗已从手术机器人，演化到5G远程手术和互联网医院齐头并进。智能政务不但覆盖全域，更是在城市交通和民生服务方面走向纵深，还连通了各个部门之间的数据，解锁了"一键办理"功能。

智能产业的新浪潮

毫无疑问，重庆已逐步成为中国智能产业新的蓝海。

前沿的技术在这里诞生，新潮的思想在这里碰撞，未来的构想在这里展开。

在这里，或许能满足人们对智能产业的一切想象。

自2017年起，重庆在制造业展开智能化改造，运用大数据智能化技术改造提升传统制造业，并大力培育大数据、人工智能、智能硬件等智能产业链。截至2020年7月底，重庆累计完成智能化改造项目2 200个，建成数字智能工厂67家、数字化车间539家。[1]

另外，重庆市着力构建"芯屏器核网"全产业链，抢抓新一轮物联网发展机遇，充分发挥平台运营和应用示范优势，努力加强MEMS传感器、通信模组设计制造等，全力打造硬件制造、运营服务和系统集成"三位一体"的产业体系。

重庆的智能产业释放了产业布局信号，引发了广泛的关注。看到了西部的广阔前景，更多的企业选择落户重庆。

与此同时，一次"孔雀西南飞"的智能产业新浪潮，也在

---

[1] 赵宇飞、伍鲲鹏，新华社，《中国西部正崛起为智能产业新高地》，2020年9月14日。

图片来源：张锦辉 / 视觉重庆
Photo by : Zhang jinhui / Visual Chongqing

市民在位于两江新区水土园区的重庆市半导体科技馆时光隧道了解半导体发展历程。

Citizens learns about the history of semiconductors in the Time Tunnel of Chongqing Semiconductor Science and Technology Museum, located in Shuitu High-tech Industrial Park of the Liangjiang New Area

更大范围内滚滚向前。

近几年，四川、贵州等西部省份的智能化步伐也在不断加快，四川省出台《四川省新一代人工智能发展实施方案》等行动计划，贵州大数据产业发展指数位列全国第三，已连续多年稳居全国前列。

拥有4亿人口、占国土面积三分之二的中国西部地区，发展潜力巨大，生产要素成本优势明显，智能产业正在崛起。

第 2 节 战略之城：
加快建设"智造重镇""智慧名城"

古之立大志者，不惟有超世之才，亦必有坚韧不拔之志。

——苏轼

《孙子兵法》讲，"道为术之灵，术为道之体；以道统术，以术得道"。

在重庆，道就是城市的战略，术就是智能产业。

从"制造重镇"到"智造重镇"，重庆开启转型进程尚不足十年。整个过程思路清晰、毫不迟疑，从物联网基地到大数据中心，从数据交互到政务上云，从名企落地到万物生长，用最短的时间，成为中国智能产业新高地。

而这一切，取决于"道"的正确。

智慧城市的战略演进

一个"善感知、会呼吸、有温度"的未来城市，到底应该走在一条什么样的路上？

在重庆之前，全球虽然也有城市在试验智慧城市之路，但都还未成型，仍在不断的探索中。重庆也一样，在摸索中不断修正自己的路。

2018 智博会在重庆盛大开幕
SCE 2018 grandly opens in Chongqing

――――――――――――――――― ······

2018年6月，在智博会发布会上，重庆发布了《重庆市以大数据智能化为引领的创新驱动发展战略行动计划（2018—2020年）》，实施大数据智能化发展战略。此次战略提出，加快推进数字产业化、产业数字化，重点围绕智能网联汽车、智能制造、智能感知、智能物联网、智能机器人、智能终端、集成电路、云计算大数据（超级计算机），人机交互等产业领域，初步构建开放协同的大数据智能化产业创新体系。

要完成这个宏大的规划，当然少不了5G。

2019年，重庆针对热门的5G技术，发布了《重庆市加快推动5G发展行动计划（2019—2022年）》，抢占5G发展先机，加快重庆5G网络部署、产业发展和商用步伐,推进"智造重镇"和"智慧名城"建设,培育经济增长新动能。同年,重庆发布的《重庆市新型智慧城市建设方案(2019—2022年)》明确提出,到2022年,建成全国一流的大数据智能化应用示范城市、城乡统筹开放发展的智慧社会样板,以新型智慧城市融合创新动力带动重庆实现新一轮跨越式发展。

依次出台的战略方案和行动计划中,重庆转型发展智能产业的行进路线一目了然,每一步都将这座城市指向"智造重镇""智慧名城"。

老工业城市的新生之路

提起中国的智能制造,人们首先想到的可能是北上广深,以及东部沿海各省,地处西南的重庆,仿佛很难引人注意。重

在探索智慧城市的道路上,重庆有自己的智慧经验
Chongqing accumulates its own experience in building smart city

庆市作为中国近代工业发展较早的城市之一，也是重要的老工业基地之一，在现代制造业升级中形成了全球最大电子信息产业集群和中国领先的汽车产业集群，成为全球最大的笔记本电脑生产基地和全球第二大的手机生产基地。在装备制造、综合化工、材料、能源和消费品制造等领域，重庆同样有着自己的产业优势，产业集群规模也是千亿级别。

老工业城市发展智能制造，早已准备好了。

重庆地处长江上游，千百年来天然形成了包容与开放的城市文化，在这种城市文化氛围中，重庆人的骨子里也滋生着一种"天下无事不可为"的基因。当年从内陆的贸易之都顺利转型成重工业城市，现在从工业城市转型为智慧城市，在高屋建瓴的宏大叙事背后，重庆的城市文化、重庆人的特质，同样提供了源源不断的能量。

作为"一带一路"和长江经济带的联结点、内陆国际物流枢纽，重庆，正以一种崭新的姿态迎接着新时代的机遇和挑战。

在智慧城市的规划中，智能制造一直是关键点。各大本土制造企业，也在这场轰轰烈烈的智能制造浪潮中闪亮登场。

宗申集团是重庆摩托制造领域的龙头，也是智能化改造的先锋，旗下的工业互联网全产业链创新平台忽米网，构建了全国动摩行业首个，同时也是重庆首个工业互联网标识解析二级节点平台忽米云析，截至2020年12月3日，标识注册总量已突破400万，标识解析总量突破4000万，日均解析量达近30万，帮助重庆市标识解析量进入了全国城市第七位。[1]

在2020线上智博会"智慧重镇"展馆，忽米网把"宗申1011智能生产线"的迷你版搬到了现场，让全球观众可以看到

---

[1] 杨野，上游新闻，《忽米网CTO陈虎：工业互联网标识解析核心是为企业赋能》，2020年12月3日。

忽米网将宗申1011智能生产线场景带到2020线上智博会"智造重镇"展馆现场

Humi.com brings a mini version of "Zongshen 1011 Intelligent Production Line" to "Smart Manufacturing City" Exhibition Hall at the 2020 SCE Online

......

整个装配生产线的生产制作过程。通过标识应用，宗申1011生产线进行智能化升级改造后，宗申动力摩托车总装生产线整体效率提升了4倍，人员减少了70%，自动纠错防错能力提升了10.6倍，作业自动化率提升了10倍。宗申集团整体生产实现了设备使用率提升了25%，生产效率提升了15%，不良品率降低了20%，设备健康度提升了20%。[1]

同时，重庆的汽车制造商也在向智能化靠拢，以工业互联

---

[1] 郑三波、彭晨，上游新闻，《探馆：把宗申1011智能产线迷你版搬到现场是种什么体验？》，2020年9月14日。

网为核心，一边改造着自身的生产工厂，一边帮助产业链上的其他制造商进行智能化生产。

长安汽车就是其中的先行者，他们一方面联合重庆联通，打造基于"5G+工业互联网"的协同智造工厂，一方面构建全球网络协同设计和制造体系，贯通产业链、价值链、资源链等上万家企业，由下而上，由内而外进行智能化升级。在长安智造工厂里，人、机、料、法、环全生产要素都通过5G连接，对工厂车间内生产运行状态实时监控，实时管理人员、设备、产能、能耗、物流等信息，实现工厂透明化管理和无人化生产。

在智能化改造中，制造企业进一步成长，重庆也获得新的增长动力，行业巨头争相落户重庆。

2020年10月，吉利工业互联网全球总部正式落地重庆。以吉利三十五年制造业经验与资源为依托，吉利工业互联网平台具备强制造底蕴、多应用场景、全覆盖领域三大核心竞争力，以跨行业、跨领域、多场景的解决方案为产业转型提供平台级服务。吉利工业互联网平台通过构建集资源能效、安全可信、数据智能、智能物联于一体的数字化基座，让企业拥有数字化转型的一体化基础能力，支撑工厂数字化、数字化运营、C2M柔性定制、智慧出行、双碳管理五大解决方案，推动产业高质量发展，为社会创造高品质生活。

未来城市已在路上

未来已来，重庆已经在路上，即将看到智慧城市成型的曙光。

在新的一年，重庆也有新的规划，在2021年的智能化改造升级规划中，重庆市拟定了推动实施1 250个智能化改造项

吉利新能源汽车换点示范站
Battery-swap Demonstration Station for Geely Alternative Fuel Vehicles

......

目的目标。[1]包括重庆市辖区内实施数字化装备普及、数字化车间和智能工厂、工业互联网平台建设和"上云上平台"、智能制造新模式应用等项目。[2]

智能化改造有利于提升工业企业智能化水平、推动制造业智能化转型升级，对企业降本增效、产品迭代升级，进而实现工业高质量发展，有着至关重要的作用。在5G基站建设、特高压、

---

[1] 重庆经信委，《关于开展2021年重庆市智能化改造项目认定工作的通知》，2021年2月23日。

[2] 刘翰书，上游新闻·重庆晨报，《今年重庆将推动1 250个智能化改造项目 快来看看目标任务分解标准》，2021年2月23日。

城际高速铁路和城市轨道交通、大数据中心、人工智能、工业互联网等领域，重庆也正通过筑牢"硬件基础"给创新加把劲，以便抓住新机遇、拓展新空间、培育新动能，为重庆经济发展插上"翅膀"。目前，重庆已建成5G基站4.2万个，实现全市所有区县重点区域5G网络全覆盖。[1]

在此基础上，为有力支撑5G等新型基础设施建设，重庆市通信管理局于2020年编制发布了《重庆市国土空间规划通信专业规划—5G专项规划》，2020—2025年，重庆将围绕5G通信网络，按照全市"一区两群"整体空间布局，结合产业发展，至规划期末，建成超高速、大容量、智能化、泛在感知的万物智联通信基础设施，实现"规划一张图、建设一盘棋、发展一体化"，在全市范围内建成15万个5G基站，达成5G整体服务水平全球领先的目标。[2]

战略规划层层推进，智能化改造密集实践，今日重庆已显露出未来城市的雏形。

而未来，正等着我们去敲门。

---

[1] 黄光红，重庆日报，《重庆建成4.2万个5G基站 所有区县重点区域实现全覆盖》，2020年7月3日。

[2] 李舒，上游新闻·重庆晨报，《重庆发布5G新基建规划 2025年前将建成15万个5G基站》，2020年5月15日。

第 3 节　实践之城：
从名企落地到万物生长

种一棵树最好的时间是十年前，其次是现在。

——丹比萨·莫约

十一年前，重庆建设总部城的时候，根本没想到这棵树能长得如此枝繁叶茂。

多家世界 500 强公司的总部相继落地重庆，为重庆带来前沿的技术，新潮的理念，以及不断试错后得出的宝贵经验。

这棵大树下的绿荫，也带来一片阴凉。本土企业站在巨人的肩膀上，不断创新，驶入发展快车道。

知名企业落地生根带来集群效应

十年之前，中国正在移动互联网的蓝海里驰骋，"人工智能"还是相对遥远的词汇，而在十年间，中国人工智能产业在国家政策与企业创新等各方推动下，进入爆发式增长阶段。

这个过程中，知名企业落地生根带来的集群效应，滚雪球般越滚越大，给重庆带来新的机会。

现在，阿里巴巴、腾讯、英特尔、小米、百度、吉利等企业都纷纷在重庆落户，已经与重庆展开了产、学、研、用等多个领域的深度合作，实现互利共赢，助力重庆智能化创新快速发展。

特别是从2018年开始，重庆市连续三年成功举办智博会，借助全球智能产业发展智慧，汇集全国智能产业资源落地，一步一个脚印，无缝衔接，节节攀升，向外界展现了强大的转型和加速发展的信心。

在此过程中，重庆市与腾讯、华为、阿里巴巴、浪潮等众多数字经济龙头企业，在众多领域展开了深度合作。

2020年3月到4月，短短一个月内，更是有八家知名企业纷纷落户重庆，智能产业的创新土壤成为最核心的驱动力。

汽车是重庆重要的工业制造品类，汽车知名企业多年以前就已布局重庆，眼下正是加大筹码的时刻。2020年3月底，福特汽车旗下林肯品牌首款国产车型"冒险家"在渝下线。4月23日，两江新区管委会、东风公司、小康股份三方共同签署《关于共建中高端新能源汽车项目的协议》，促进重庆金康新能源汽车有限公司的中高端新能源汽车项目发展，在2025年前将之打造成为年产15万辆、产值在300亿元以上的中高端新能源汽车标杆企业。[1] 4月2日，复星集团深化与重庆的战略合作，将参与部分重大交通项目。

腾讯、阿里巴巴、浪潮、华为也再次与重庆加深合作，在人工智能、5G、研究院等方面纷纷加码。阿里将在重庆知名产业带区域建立20个产业直播基地，合力推进"C2M超级工厂

---

[1] 严薇，重庆商报，《新能源汽车迎来"至暗时刻"？车企在渝暗战升级》，2020年7月21日。

计划",打造20个销售过亿的工厂[1]。腾讯将利用腾讯游戏光子工作室群等流量优势,提高彭水旅游商品及周边农特产品附加值,助力农特副及新文创周边产品销售。华为将与重庆联合打造未来智能汽车科技城、智能超算中心、5G工业领域联合实验室,在数据中心、物联网、区块链、工业互联网、新型智慧城市、智慧水利、智慧终端、5G、智慧园区、机器视觉、人才培养等领域展开全面深度合作。

这些知名企业也成为重庆智能产业的风向标,引领重庆产业创新。

本土创新万物生长

一直以来,重庆都足够包容,各种天马行空的想象,都能在这里找到落地生根的土壤。

据重庆市统计局发布的《2020年重庆市国民经济和社会发展统计公报》显示,目前重庆集聚大数据智能化企业7 000余家,实施智能化改造项目2 780个,数字经济增加值占地区生产总值比重达到25.5%。

目前,重庆人工智能产业正处于加速发展的黄金阶段,重庆本土企业经过沉淀和积累,正奋力扛起重庆人工智能的大旗。2020年3月,重庆长安汽车、马上消费三家企业凭借在各领域的创新性项目入围国家工信部"新一代人工智能产业创新重点单位"。[2]

长安汽车的"智能驾驶舱域控制系统"项目,基于人工智能、

---

[1] 韩梦霖,新华网,《阿里巴巴将在重庆建立20个产地直播基地助渝货外销》,2020年4月8日。

[2] 谢力、王丹,两江新闻,《重庆两江新区多家企业跻身新一代人工智能产业创新"国家队"》,2020年3月1日。

大数据、云计算、工业互联网等多学科交叉上位机系统,包括"数据平台+计算平台+3个开放的容器",实现千人千面与服务找人,从而最终构建智能全场景出行生态,满足用户个性化体验的需求。在智慧交通领域,为同行提供了参考。

马上消费作为一家科技驱动的金融机构,目前累计申请专利数量已超过220项,自主研发核心系统已突破700套。[1] 其中,"基于情感光谱与多语境感知的智能文本与语音客服机器人平台"项目便是重要成果之一,该项目采用前沿的人机对话系统架构,以深度学习、迁移学习、强化学习、统计学习等核心算法为基础,实现了意图识别、意图预测、用户画像等核心模块,构建出能够在多终端、多渠道24小时不间断随时响应客户问题精准应答的智能客服机器人。

在长安汽车、马上消费等本土企业的示范效应之下,重庆本土创新力量也涌现出万物生长之势,忽米网、摇橹船科技、允升科技等一批工业互联网的创新先锋陆续登场,成为重庆拥抱智能时代的新锐力量。

目标指向智能时代的头等舱

伴随着智博会,重庆智能产业的发展,从知名企业落地的集群效应到万物生长,只用了三年。

2020线上智博会上,观点碰撞出的思想之花,为重庆,乃至中国和世界在未来的智能制造、5G科技、工业互联网、人工智能等大数据智能化领域,汇聚强大智力财富,勾勒智能科技世界发展新蓝图,发出重庆声音、贡献重庆力量。

---

[1]谢力、王丹,两江新闻,《重庆两江新区多家企业跻身新一代人工智能产业创新"国家队"》,2020年3月1日。

第 5 章 重庆：智能时代的关键见证

智慧名城重庆，充满想象空间
Smart Chongqing is full of imagination

———————————— ......

腾讯公司首席运营官任宇昕认为，"随着数字世界与实体世界的融合，生产生活都在被数字化重塑"。[1]

目前，重庆已初步形成集研发、整机制造、系统集成、零部件配套、应用服务于一体的全产业体系，在物联网、智能机

---

[1]孙磊，上游新闻·重庆商报，《腾讯任宇昕：纯粹的"线下生活"和"传统产业"将不存在》，2020年9月15日。

器人、云计算大数据、人工智能等领域，都颇有成果。

并且集中力量，大力发展产业互联网，推动传统产业智能化改造，加快形成"芯屏器核网"全产业链和"云联数算用"全要素群的发展优势，塑造"住业游乐购"全场景，引导各类企业上云、用数、赋智，增强经济社会发展的新动力源。

重庆智能产业发展战略目标清晰，建设"智造重镇""智慧名城"步履不停，目标指向了智能时代的头等舱。

进入头等舱的主要筹码，则是重庆智能产业的 7 000 多家企业，这些企业形成的规模效应，正吸引着全球的目光。而目光的最深处，是对智能时代的渴望与追求。

第 4 节 未来之城：
打造智能产业时代的全球智慧名城

教诲是条漫长的道路，榜样是条捷径。

——塞内加

一座城市，或多或少，或深或浅，可以被打上多个标签和印记。

重庆当然也是如此，火锅、汽车、美女、8D 城市，甚至是地铁穿楼，不同的人会看到重庆城的不同角度，在心目中为这座城打下的标签也各不相同。

然而，无论是谁，重新打量重庆的时候，都会看到一个既新鲜又深刻的"智慧城市"印记。

重庆，已在智慧城市的成长之路上，留给世界一个清晰的印象。

重庆,既是山水魅力之都,更是"智慧"发展之城
Chongqing, the city of natural beauty and the city of "smart" development

重庆,投身全球智能化进程

时间回溯到 1956 年,达特茅斯学院举办的一次会议上,一群科学家聚集在一起头脑风暴,计算机学家约翰·麦卡锡提出了"人工智能"一词,人工智能正式诞生。随着人工智能的发展,人们对未来有了新的展望和期待,全球智能化已箭在弦上,先行者们投入其中,向未来奔去。

之前,重庆一直以工业城市的形象现身,看似和人工智能、智能城市这些前沿的词汇毫不相及。

可人世间最快意的事,永远是万丈高楼平地起。

重庆做到了,用最快的速度,提出了建设"智造重镇""智慧名城"的战略蓝图,而每年举办的智博会既是其重要的推手,也是其耀眼的成果。

虽然 2020 线上智博会因为新冠肺炎疫情改成线上展会,但依然深刻影响重庆乃至中国的新一轮科技革命和产业变革,从生产车间到建设工地、从物流配送到智慧小区……

重庆正加快智博会成果的落地见效。

如今在重庆,大多数规模以上制造企业最繁重繁琐的活儿都交给了机器人,"远程会诊"触屏操作代替了人工流水作业,智能生产线、智能物流车等设备让生产效率更高、成本更低。依托大数据智能化,重庆经济驶入高质量发展"快车道"。

去年,重庆智能产业交出了一份令人满意的成绩单。

2020 年,"芯屏器核网"智能产业架构日趋完善,集成电路、新能源及智能网联汽车、智能手机等一批重点项目相继实施,智能产业销售收入增长 12.8%。数字化车间和智能工厂建设加快推进,实施 1 297 项智能化改造项目,认定 210 个市级示范性数字化车间和智能工厂。截至 5 月 24 日,国家顶级节点(重庆)标识注册总量达 6.31 亿,累计解析 3.34 亿次,接入二级

2020线上智博会前期活动"集成电路产学研政协同创新交流会"

Collaborative Innovation Exchange Meeting of Enterprises, Universities, Research Institutions and Governments at the early stage of the 2020 SCE Online

节点19个,接入企业节点1 009家。[1]

与此同时,重庆的5G网络、工业互联网、大数据中心等新基建项目也加速上马,加速推动新旧动能转换,成为全球智能产业中一道亮丽的风景线。

用工业互联网和大数据,回应智能时代

重庆智能产业的成果,吸引了全球的目光。

所有的发明与创造,都是为了更好地推进我们的生活。人

---

[1] 重庆市经济与信息化委员会,《2020年重庆工业和信息化发展概况》,2021年4月6日。

们想要光，于是有了电灯；人们想要越过天涯的遥远，于是有了火车、飞机。这一次也不例外，人们想要更便捷的生活，于是有了各种智能产品。

提起智能产品，多数人更关注产品本身的智能属性，而忽略了产品背后生产制造过程的智能程度。而在一个全新的智能产业时代，消费者能体验到什么样的智能产品，智能制造往往发挥的作用更大。

在长安汽车渝北工厂里，数百台机械臂不停舞动，生产线上自动涂胶机器、自动装备器、图像精准定位仪等智能设备应有尽有，实现了无人化操作。重庆金康赛力斯两江智能工厂里，1 000多台机器人替代了传统生产线上的人头攒动，生产线关键工序全部实现智能化，全过程只需几位技术人员通过屏幕进行操控。

"智能改造"已在重庆制造业中全面开花，如今，全球每3台笔记本电脑、每10台手机，就有一台是"重庆造"[1]，重庆引以为傲的汽车行业，也逐步转型为如今的"智细精"。

制造业在重庆步步提升，与此紧密相连的工业互联网也走上新的台阶。重庆智能工厂、智能车间、工业互联网发展大步向前。

2018年，腾讯西部云计算数据中心一期已经正式建成并启动试运营，是腾讯布局的自建大型数据中心集群之一，已经成为腾讯在西南地区重要的数据中心和网络中心。目前，腾讯西部云计算数据中心二期已建成大半，建成后，能有效带动腾讯西部云计算数据中心形成20余万台服务器计算能力，并成为中国西部最大的单体数据中心。

---

[1] 郑三波，上游新闻·重庆商报，《全球每10台手机、每3台电脑就有一台"重庆造"》，2020年5月22日。

2020年6月，《重庆市新型基础设施重大项目建设行动方案》出台[1]，明确提出3年里，重庆将总投资近4 000亿元，滚动实施和储备375个新基建重大项目。重庆邮电大学正在建设的大数据智能化实验场携手浪潮集团落地了人工智能创新平台，打造面向大数据的技术研发和试验验证环境，其中，"算力"是核心底座。

2020年年底，重庆市大数据成果转移转化中心正式授牌成立，将围绕大数据产业发展方向和大数据企业需求，搭建资源共享平台，遴选一批技术成熟度较高、市场前景较好的科技成果，建立高价值成果库；开展线下和线上相结合的科技成果发布、技术供需对接，促成国内外大数据应用成果在渝转化落地。

智慧名城的终极构想

在边缘机房、云计算中心、政务中心、物联网基地、智能车间里，重庆智慧城市已筑好地基，整个城市从内部发出智能改造的轰鸣声。

而"智慧名城"的构想已静静矗立在礼嘉智慧公园，审视着未来的每一步。

礼嘉智慧公园于2019年8月正式开园，在2020线上智博会期间，20余个智能制造场景和10余种智能产品汇聚礼嘉智慧公园，展示重庆这座"智慧名城"的风采，搭建起永不落幕的智慧舞台。

公园的二期工程——湖畔里数字中心再次惊艳世界，更加先进的科技、更加智能的园区运营、更加丰富的智慧生活体验和更多的智能机器人，让这棵"智慧之树"更加茁壮。

---

[1] 张瀚祥，上游新闻·重庆晨报，《重庆计划到2022年累计投入3983亿元布局"新基建"》，2020年6月21日。

礼嘉智慧公园已经成了重庆智慧名城的地标
Lijia Smart Park becomes a landmark of Chongqing

湖畔里数字体验中心的井盖、路灯、摄像头、环保检测、停车场等设备和场地,均实现物联感知和智能管理,充分展现万物互联。

5G 网络在公园实现全覆盖,重点区域还建设了第 6 代 Wi-Fi 网络,成为万物互联的前哨所。

满园的文化与科技,让礼嘉智慧公园成为网红打卡地。

在"智慧交通"体验区,游客坐在未来汽车驾驶舱内,就能亲身体验自动驾驶;在"智慧文旅"区,游客戴上 MR 眼镜,说出目的地后,李子坝穿楼轻轨、长江索道等热门景点就会迎面而来;在"智慧医疗"区,则向游客展示了 AI 未来诊疗室,可提供贯穿患者全生命周期的智慧医疗解决方案。

从踏入公园开始，未来就已经到来。我们设想的所有关于未来的场景，都能在公园里看到相应的影子。这给全球智能化以信心，给智能时代以指引。按图索骥以后，未来是可以预知的，是一连串的问号变成惊叹号的过程，是期待与惊喜的相逢。

而这，正是一座创新之城探索一个伟大时代的意义所在。

后记
智能时代的年度印记

以大数据、5G、人工智能与物联网等为代表的新一代信息技术，正在交叉与叠加，同时与实体经济深度融合，由此引发人类社会爆发式变革。在"发展数字经济，推进数字产业化和产业数字化"的国家战略下，重庆市明确提出了建设"智造重镇""智慧名城"的目标，并规划了"芯屏器核网""云联数算用"与"住业游乐购"的实施路径。永久落户重庆，每年一度的智博会，以"智能化：为经济赋能，为生活添彩"为题，已经成为具有国际影响力的重要盛会。

作为伴智博会而生的系列图书，"解码智能时代丛书"自2020年出版以来广受好评。它的文字生动可读、通俗易懂，既总结了智博会的交流成果，又展望了全球智能产业的发展趋势。

为了呼应智能时代发展的新趋势，展示智能产业取得的新成果，同时也为广大群众打造一套了解智能时代、融入智能时代的优秀科普读物，中共重庆市委宣传部决定持续性地出版"解码智能时代丛书"。中共重庆市委常委、宣传部部长张鸣同志对"丛书"进行了全面指导，明确提出：要以国际化的标准，将"解码智能时代丛书"打造为智博会的一张文化名片，以及在智能化领域具有重要影响力的系列读物。

"解码智能时代丛书（2021）"立足于重庆市"智造重镇""智慧名城"建设总体战略，围绕2020线上智博会的交流成果、全球智能产业理论和实践的最新探索组织编写。"丛书"策划内容共3种，其中《解码智能时代2021：从中国国际智能产业博览会瞭望全球智能产业》以图文并茂的形式呈现了2020线上智博会的丰硕成果和智能产业的发展现状及趋势；《解码智能时代2021：来自未来的数智图谱》从"芯屏器核网""云联数算用"与"住业游乐购"的角度，解读了重庆在建设"智造重镇""智慧名城"方面的最新实践；《解码智能时代2021：前沿趋势10人谈》涵盖了10个话题，访谈了10位来自全球理论研究和智能产业领域的代表人物，其中既有院士、教授，也有知名企业家。"丛书"构成了一个立体、多维、丰富的观察体系，在2021智博会召开之际，记录了全球智能产业的新成果，展望了全球智能时代变革的新趋势。同时，为了讲好中国故事，并与全世界的读者分享智能产业领域的中国实践，相关创作内容同时配有英文译本。

"丛书"的组织策划、调研写作及编辑出版是一个庞大的系统工程，由中共重庆市委宣传部策划组织，并在重庆市经信委、重庆市大数据应用发展管理局等部门的配合下，由黄桷树财经等多个专业团队创作，重庆邮电大学MTI团队翻译，重庆大学出版社出版。

在整个写作及出版过程中，中共重庆市委宣传部常务副部长曹清尧同志，市委宣传部副部长、市新闻出版局局长李鹏同志对"丛书"的写作和出版工作做了具体安排部署，市委宣传部出版处统筹多个专业团队紧密协同，对"丛书"的策划创意和内容质量进行总体审核把关，推动完成了"丛书"的编写及出版；重庆市大数据应用发展管理局副局长杨帆同志、重庆市经信委总工程师匡建同志对创作工作进行了专业指导；智博会秘书处、重庆市九龙坡区融媒体中心、重庆市两江新区融媒体中心、重庆大数据人工智能创新中心与公共大数据安全技术重点实验室对创作工作进行了大力支持。重庆大学出版社特别邀请智博会秘书处何永红主任、重庆市经信委刘雪梅处长、重庆市大数据应用发展管理局法规标准处杜杰处长以及重庆大学的李珩博士、李秀华博士对"丛书"进行了审读，重庆大学出版社组织了多名资深编辑对书稿进行了字斟句酌的打磨，从而确保内容的科学性、可读性及准确性。

如果站在智能时代历史进程的维度上，我们希望"解码智能时代丛书"能够以年度为单位，记录与展望智能化究竟如何为经济赋能、为生活添彩，记录与展望"数字产业化、产业数字化"的实践过程，记录与展望人类文明史上这场伟大而深刻的变革。这样的记录与展望，这样的智能时代年度印记，是有历史意义的。

谨此，致敬中国国际智能产业博览会，并对所有促成本书立项、提供写作素材、执笔书稿编写与翻译、参与本书审订、帮助本书出版的单位与个人，对接受写作团队采访的专家，致以深深的谢意。

编写组

2021 年 7 月

DECRYPTING THE INTELLIGENT ERA 2021

Overlook the Global Intelligent Industry from Smart China Expo

Chongqing University Press

Preface

This book is not a simple record of the China International Smart Industry Expo (hereinafter referred to as "Smart China Expo").

It takes the Smart China Expo as a window to guide readers to experience how the digital economy represented by intelligence has changed the world, how digital is industrialized and how the industry is digitalized. It tells what is happening now and what will happen in the future. It is a gymnastics of minds, and a feast of ideas.

The book is divided into five parts, which includes "Grand Event" , "Trends , " "Practice , " "Achievements" and "Chongqing."

"Grand Event": A record and introduction to the 2020 Smart China Expo Online. It highlights the record of innovative ideas and shows the intellectual evolution of the global elites in the direction of intelligence.

"Trends": After recording thoughts, there are ideal and practical developments, and the trend is "ideal." What is the direction of the smart industry? The future is here, and the general trend is already clear.

"Practice": "Trends" follow the the blueprint, and "practices" are concretely implemented and applied in various scenarios, such as manufacturing, agriculture, urban construction, and public health.

"Achievements": Going back to the readers and reviewing the achievements of artificial intelligence (AI), such as smart materials, industrial interconnection, autonomous driving, smart cultural tourism, smart epidemic prevention, smart finance and smart government. All achievements are no longer in our imagination, but are perceivable realities.

"Chongqing": Standing in a brand-new era, and reviewing from the perspectives of imagination, strategy, practice, and the future, this book explores this "smart manufacturing town" and this "smart city", where it comes from and where it goes.

The most significant value of this book is to guide readers to witness history.

The inventor Kurzweil proposed the Law of Accelerated Returns: Technological progress is advancing rapidly at an exponential rate. All people in this era are on the cusp of accelerating change, and the speeding development we are experiencing exceeds any moment in human history.

However, most of us, perhaps, have not felt it particularly intuitively. Like a fish in a big river, it may not necessarily appreciate how turbulent the river is. And this book is to help the fish jump out of the water and have a broader view of the turbulent and great era.

Every reader bears witness to this great age.

It has taken millions of years for humans to evolve from hominids to the present, and the pace of evolution has been accelerating, as measured by the different stages of civilization.

The history of hunting civilization is measured in tens of thousands of years. Agricultural civilization lasted for thousands of years, while industrial civilization is only a few hundred years old. Computer and AI civilizations, which began decades ago, are currently undergoing rapid development and evolution. Any completely new stage in human civilization takes a dramatic decrease in time, while the impetus for human society increases exponentially.

And now in the age of intelligence, each of us will witness far more dramatic changes than history.

How to read this book?

First, read this book in the specific context of 2020.

The Smart China Expo has been permanently settled in Chongqing, and it will be an eternal topic in Chongqing once a year. But the scene will be different every year.

In 2020, COVID-19 pandemic has posed challenges for the whole world. The more challenges we face, the more power of human progress we must demonstrate to make the world better and safer through technology.

This book records the 2020 Smart China Expo Online. It was a year full of challenges, and the challenges have brought about many innovations, such as the integration of online and offline, reality and virtual, and the application of modern information technologies of VR, AR, and digital twins.

Second, it is necessary to read the cases as well as to feel the cognitive observation in the book.

This book deals with AI, involving both the "causes" of technological developments and the "achievements" of industrial trends. But the application cases in the book focus more on the present, unfolding the changes that are taking place in the vast process of the global intelligent industry to the readers.

This book also focuses on interspersing itself with the cognitive changes in the intelligent age and inspiring readers to think deeply.

With the deepening of technology, the degree of intelligence will become higher and higher. While enjoying the convenience brought by technology, we may also bear the pressures brought by it. But no one can stop the pace of technological changes. Only by discovering the world, discovering ourselves, and keeping up with trends can we resonate with the times.

CONTENTS

Chapter 1

Grand Event:
Evolution of the Thinking of Global Elites

002 / Section I Practice:
From Looking Up at the Starry Sky to Being Down to the Earth

005 / Section II Acceleration:
From Sketching Lines to Splashing Ink

009 / Section III Change:
Artificial Intelligence in the First Scene of All Changes

Chapter 2

Trends:
Key Navigation in the Intelligent Industry

014 / Section I   Jump–Start 5G,
Key Layouts Affecting the Global Intelligent Industry

019 / Section II   Foresee the Next 30 Years of Intelligent Industry from New Energy and Carbon Neutrality

024 / Section III   Industry Without Boundaries, Artificial Intelligence Throughout All Scenarios

029 / Section IV   Digital Life and Artificial Intelligence Redefine Social Relationships

034 / Section V   Smart Dividend:
China Unleashes New Growth Momentum for the World

Chapter 3

Practice:
The First Site of Intelligent Industry

040 / Section I   Smart Manufacturing, the Transition of the Industrial Era or the Budding of the Intelligent Era?

045 / Section II   Smart Agriculture, Redefining the Production Relationship Between People and Agriculture

049 / Section III   Smart Cities, the Fastest Transformation in the History of Urban Development

054 / Section IV   In the Innovation Scene, Artificial Intelligence Transforms All Industries

059 / Section V   China Adopts AI Practices to Fight Against the Epidemic Under the Pandemic Dilemma

Chapter 4

Achievements:
Cutting-Edge Innovations in Technological Intelligence

066 / Section I   The Convergence of Black Technologies, Overlook the Future of the Times from the Smart China Expo

068 / Section II   From Smart Materials to Autonomous Driving, Manufacturing Is Changing at an Accelerated Pace

076 / Section III   New World at the Smart China Expo Empowers Life to Leap "Intelligently"

083 / Section IV   In the Intelligent Financial Transformation, Technology Brings the "Turbo-Charging" Effects

088 / Section V   The Convenience of Life Starts from the Smart Government Services

Chapter 5

Chongqing:
A Key Witness of the Intelligence Era

094 / Section I   A City of Imagination: Chongqing Is Committed to the New Trend of Intelligent Industry

099 / Section II   A City of Strategy: From a Manufacturing Town to a Smart City

104 / Section III   A City of Practice: From the Landing of Famous Enterprises to Driving the Rapid Development of Various Industries

108 / Section IV   A City of the Future: A Global Smart City in the Era of Intelligent Industry

Chapter 1

Grand Event: Evolution of the Thinking of Global Elites

In the past year, we have witnessed a round of unprecedented test and risks, a reckless journey of the times, a city of wisdom with unwavering determination, an industrial gathering that never ends. This year, for every country, every city, every company, every individual has a different meaning from the past. From encompassing and spreading, shattering and reshaping, discarding and accepting, to parting and reuniting, every moment is staging conflicts, and every moment is trying to integrate.

The whole world, everything that was once certain seems to be no longer certain, while the wheels of the intelligent era are still rolling forward, the development of artificial intelligence, instead, has become the greatest certainty in the global uncertainty. It is in this situation that China International Smart Industry Expo (hereinafter referred to as "Smart China Expo") once again opened the curtain from the city of Chongqing, China.

Of course, there are changes. From looking up at the starry sky to being down to the earth, from sketching line drawings to splashing ink, artificial intelligence is no longer limited to the imagination, but a more pragmatic stance, being on the first site of all the changes.

Section I

Practice:
From Looking Up at the Starry Sky to Being Down to the Earth

The greatest worship we can express to the truth is to perform it down-to-earth.

— Ralph Waldo Emerson

No one could have expected that 2020 would be such a sudden year, still less could anyone have predicted that 2020 would be such a long year.

The COVID-19 epidemic has cast a shadow over 2020, affecting almost all major activities, events, exhibitions, forums and summits.

It has been a year without sporting events, and most major global sporting events have been postponed. Olympic Games, one of the world's largest and most influential human group activities, has been held every four years since 1896. It was only interrupted for three times due to the two World Wars over more than 100 years, and the 2020 Tokyo Olympics became the first postponed event. The five major European football leagues (La Liga, English Premier League, Serie A, Bundesliga, French Ligue 1) and the European Cup were postponed indefinitely. FIBA U16 Asian Championships & FIBA U16 Women's Asian Championships decided to cancel the game ...

It has been a year without industrial exhibitions. Dozens of international industrial exhibitions around the world have been canceled or postponed. The Taipei Game Show in Taiwan, China, Apparel Asia in Berlin, Germany, Art Basel in Hong Kong, China, and the Tokyo International Retail Application Fair in Tokyo, Japan ... If you were to list them all, the list could be a long one.

It has been a year of limited scientific and technological exchanges. Many top summits in the global scientific and technological field have been completely canceled by the organizers, including Mobile World Congress (MWC), Game Developers Conference (GDC), Google I/O Global Developers Conference, the Optical Fiber Communication Conference and Exhibition (OFC), LED NEXT STAGE in Tokyo, Japan ...

Wait a minute, is waiting the only thing the whole world can do under the plight of the epidemic?

Even in the coldest of harsh winter, there are shoots that break through the tundra. The epidemic is a frozen moment for the world, and the entire world needs a force that can empower it to explore and move forward.

Chapter 1 Grand Event: Evolution of the Thinking of Global Elites

The initial response of the top summits, like Microsoft Build 2020 Conference, Apple WWDC 2020 Developer Conference, Google Cloud Next 2020 Summit, NVIDIA GTC-GPU Technology Conference 2020, Adobe 2020 Digital Experience Summit, 2020 Tencent Global Digital Ecological Conference, 2020 Alibaba Cloud Online Summit, 2020 Baidu Yunzhi Summit, Huawei 2020 Win-Win Global Online Summit, and other top summits of leading global technology companies, was to postpone the summits, but later they have switched to online summits.

Yes, the world has its realistic problems, and technology has its innovative solutions.

The COVID-19 pandemic has created obstacles to industrial exhibitions and interactions in various fields around the world. However, the intelligent industry has its own smart way to move forward.

While the world's leading technology companies are preparing for the online event, Chongqing officially issued a global invitation to rally around the intelligent era, and the annual intelligent industry event was still held as promised.

On September 15, 2020, the 2020 China International Smart Industry Expo Online (Hereinafter referred to as "2020 Smart China Expo Online") officially opened in Chongqing, and visitors could browse the exhibition online from anywhere in the world. The 2020 Smart China Expo Online has broken through the "face-to-face" geographical restrictions, realized "screen-to-screen" communication and interaction, and opened up an "end-to-end" global gathering.

Under this particular global background, it was a surprise to all countries around the world that Chongqing city could successfully host such a high-profile international exhibition for the intelligent industry. Singapore's Minister of Human Resources and Second Minister of Home Affairs, Yang Liming, could not help but exclaim at the opening ceremony, "At a time when the COVID-19 pandemic continues to spread, Chongqing has overcome many difficulties and challenges to hold this event, which fully proves Chongqing's ability of innovation and application of digital technology."

The more in the midst of nature's challenges, the more it can reflect the solidarity of the global community of shared future. The 2020 Smart China Expo Online attracted 443 renowned experts and industry elites to collide ideas and exchange achievements in 41 forums centering around frontier hot topics such as artificial intelligence, 5G, blockchain, Industrial Internet. It also attracted 551 domestic and foreign online exhibitors to release intelligent products and innovative achievements. At the expo's online investment promotion activity, a total of 71 projects with an investment of 271.2 billion were signed. The online exhibition halls have been viewed more than 19 million visitors.

The grand scale remained great, and the atmosphere was even better than before.

The sparks of the wise men's thoughts collided here again; the cutting-edge technologies met here once again; and the up-to-date applications were introduced here once again. Although the global level of interactive communication in the era of intelligent industry was temporarily stagnated due to the global COVID-19 epidemic, it was unimpeded by the convening of the 2020 Smart China Expo Online.

As an outpost to explore the intelligent era, the 2020 Smart China Expo Online has sounded the clarion call for progress. Within Chongqing, an intelligent transformation is being launched. Every thing, from the top-level framework of a smart city to the intelligent management of a street lamp, is being incorporated a new gene belonging to the intelligent era.

The Liangjiang Cloud Computing Center is rising, giving the whole city a smart brain. Lijia Smart Park is taking a shape, combining the scenes in science fiction movies with reality. 5G base stations have covered the city's streets and alleys. The Internet of Everything (IoE) becomes possible.

In Chongqing Lijia Smart Park, children could shuttle between various black technologies, from AR band to holographic projection, from dynamic water screen to smart soccer park, from 5G bike to smart robot...

The Smart China Expo was here as scheduled, and guests from all over the world exchanged new ideas. Chongqing welcomed the guests with a new look.

The smart city presents new smart practices. New technologies, new businesses and new models continue to break conventions and knock on the door of the future.

In an ideal smart world, every time the global intelligent industry meets in Chongqing and outlines the drafts to the blueprints, the structure goes from the divided to the whole, and the grand blueprint becomes clearer and clearer.

From looking up at the starry sky to being down to the earth, from sketching lines to splashing ink, Chongqing has held three sessions of Smart China Expo for three consecutive years. For the global intelligent industry, Chongqing has been keeping in a low profile and looking into the future with careful preparations.

Section II

Acceleration:
From Sketching Lines to Splashing Ink

The poem should be painstakingly crafted, with emphasis on variation, and the brush should be bold and vigorous.

— Dai Fugu

Looking back on all the influential creative moments in human history, we can see that the beginnings were always hazy and rough sketches, and the flourishing moments were filled with colorful splashes of ink.

From the initial sketch to the subsequent re-coloring, the process was not a quick fix, but full of countless inspirations and explorations, destruction and construction, creation and reshaping. Being in it, all the aspirations for beauty were extraordinarily heartwarming.

Birth of the Sketching Lines

People always had a vague sense of the future.

It is like doing a question by checking the required options first and then verifying the others one by one.

Based on the development of China's e-commerce and Mobile Internet and the advantages of Chongqing's strong notebook computer and mobile phone manufacturing, Chongqing made the options of mobile communications, cloud computing and the Internet of Things (IoT) for the future.

Whether artificial intelligence dominates everything, robots become our faithful companions, or everything can read human minds through brain waves, mobile communication will be the foundation of everything. Data and computing power will become the new means of production and the IoT will allow devices to become more sociable.

In January 2010, China's Ministry of Science and Technology officially recognized Nan'an District of Chongqing as "National Mobile Communication High-tech Industrialization Base." In December 2010, China's Ministry of Industry and Information Technology officially recognized Nan'an District as "National Demonstration Base of New Industrialization Industry" and Nan'an District becomes a key area for the national development of mobile communication and IoT.

On April 6, 2011, the construction of Liangjiang International Cloud Computing Center and China International E-Commerce Center Chongqing Data Industrial Park was started in Shuitu Hi-Tech Industrial Park of Liangjiang New Area, which marked the extension of Chongqing's telecommunication industry from terminal products to communication field and prepared to build the largest offshore data processing center in China.

With mobile communication, cloud computing and IoT as the backbone, the offshore data processing center will explore the development direction and application scenarios of the new generation of information technology.

If these plans fall on draft paper, there are only a few strokes. However, it was just a few simple strokes that let Chongqing feel the pulse of the future era ahead of time.

After becoming a national smart city pilot, Chongqing's sketches become line drawings, and subdivided segments, such as smart society, smart governance, smart home, smart business district and smart transportation have risen up to outline the complete image of the smart city.

Located in western China, Chongqing is not the heartland of the intelligent industry, but it has a stronger determination to embrace the future-oriented intelligent era.

Cities are accelerating forward in the intelligent process, and the growth of innovative enterprises has also embarked on the high-speed train of intelligent industry.

With the new thinking and technology brought by the landed giants, Chongqing has also nurtured local AI unicorn enterprises. The industry gathering effect is emerging, and the city has been revitalized from within with a new glow.

Acceleration of the Layout of Smart Cities

A city with a sketch line of intelligent industry will have a deeper insight into the future blueprint of the whole world. Inviting the industry creators of the whole world together will provide a stage for concentrated creation of the global intelligent industry's picture scroll, and also precipitate a thick and colorful practice for the chapter of Chongqing's intelligent industry.

The Smart China Expo held in Chongqing is the first scene where the global intelligent industry splashes ink.

The 2020 Smart China Expo Online broke the limitation of time and space. With both online exhibitors and offline experiencers, and real and virtual exhibitions, this expo enriched and diversified the display contents by creating a new online exhibition display platform with the help of modern information technologies, such as AR, VR and Digital Twin, and various of functions, such as exhibition hall guide, virtual explanation, 3D product interaction and data analysis.

In the annual Smart China Expo, the latest achievements in the field of global big data intelligence will converge in Chongqing.

The year 2020 was no exception. A number of new technologies, new products, new businesses, new applications and new achievements, such as silicon photonics technology, L4-level self-driving mid-bus and 8K Micro-LED, made their "world debuts" in Chongqing.

The eyes of the world were focused here, re-examining the future of Chongqing and the future of China's technology.

At the 2020 Smart China Expo Online, Robin Li, founder, chairman and CEO of Baidu Group believed that in the near future, Chongqing would become a gathering place for developers and entrepreneurs because of "intelligence." This change is due to Chongqing's "dare-to-be-the-first" boldness and courage.

The "integration of digital world and physical world" is becoming an engine of the smart economy and smart life. With the integration of the digital world and the physical world, production and life are being reshaped by digitalization.

In Chongqing, every street, every street light, every traffic light meets in the cloud. In Longhu Liangjiang Xinchen Yunding community, residents can go through the front gate more than ten meters away and use face recognition to inductively unlock the door by the App on their mobile phones. Sprinkler system will automatically spray water when the soil moisture is below the required parameters ...

According to Chongqing's planning, the construction of "smart city" is condensed into "One Center and Two Platforms," i.e., Chongqing Big Data Resource Center, Digital Chongqing Cloud Platform and Comprehensive Service Platform of Smart City. The simultaneous construction of various "smart city" scenario applications promotes the sharing and interoperability of big data in civil, government and commercial fields, and innovative applications.

In particular, the Lijia Smart Park has comprehensively simulated and showcased Chongqing's achievements in smart manufacturing, smart applications, smart medical care, new materials, science, and technology innovation—the smart robotic arm that accompanies the elderly in Tai chi, the 5G bike that travels around Chongqing in the clouds, the medical system for AR remote surgery, and the 5G-based remote driving system ...

In the splash of ink, the smart city is accelerating the layout. The achievements of intelligence are emerging in a concentrated manner.

With Innovation, the Future Is Coming

Steadfastly following the innovation and development path of digitalization, networking and intelligence is the persistence and practice of Chongqing for many years.

Based on the advantages of its manufacturing industry, Chongqing has initiated intelligent transformation of the manufacturing industry from different perspectives, promoting the transformation and upgrading of the manufacturing industry to high-end intelligent industry.

As early as in November 2017, Chongqing proposed to lead industrial transformation by upgrading with big data intelligence, promote the deep integration of the Internet, big data, artificial intelligence with the real economy, accelerate the development of the digital economy, and promote the accelerated development of the manufacturing industry to digitalization, networking and intelligence.

The *Work Report of the Chongqing Municipal People's Government in 2019* proposed that the intelligent industry should be cultivated and expanded, while promoting R&D innovation and complementing the chain into a cluster. The work report, for the first time, proposed that Chongqing should focus on the whole industry chain of "chips, LCD, smart terminals, core elements and IoT," persistently strengthen basic research, deepen cooperation between industry, academia and research, focus on high-end chips, basic software, core devices and other areas to speed up the breakthrough of a number of key and core technologies, and complement the chain into a cluster, thus accelerating the upgrading of intelligent terminal products.

The building of a "all-element cluster" with "cloud, Internet, data, algorithm and application" was first proposed in the *Work Report of the Chongqing Municipal People's Government in 2020*. Chongqing will coordinate the city's cloud service resources to build a shared "a cloud bearing" digital Chongqing cloud platform service system.

Chongqing is going to build a new generation of ubiquitously interconnected information network system, create a dedicated channel for international data, and achieve a significant improvement in the network system's "convergence" capability and international information hub status.

Chongqing is going to build the city's big data resource center with the goal of data concentration, forming the city's unified data resource system and data governance architecture.

Chongqing is planning to build the super algorithm capability with the intelligent hub as the core and the edge algorithm and AI computing as supplements, forming the algorithm center with generic technology.

Chongqing is going to accelerate the construction of typical intelligent applications with distinctive characteristics and innovation leadership, and continuously promote the innovative development of digital economy.

Focusing on diverse application scenarios, digital technology and intelligent elements are fully integrated into various fields of urban planning and construction management to create smart communities, smart governance, and smart transportation.

With everything being connected to the Internet, the virtual world and the real world will be merged into one, forming a new world.

In the new world, perhaps we don't need to speak or operate manually. Artificial intelligence can arrange everything we need through brain waves.

If you want light, there is light in the world. If you are thirsty, water is already at hand. If you are hungry, meals are already in front of you.

The boundaries of time and space will no longer exist, and the world, at the mercy of rushing thoughts in your mind and mine, will be transformed.

Section III

Change: Artificial Intelligence in the First Scene of All Changes

This great world forever, constantly change stereotypes.

— Alfred Lord Tennyson

The world is constantly changing, and everything contains the logic of self-growth.

In this era, it is not just technology that changes life, but artificial intelligence that transforms everything.

From the very beginning of its emergence, artificial intelligence has assumed the human imagination of the intelligent era, and other technologies have taken a qualitative leap after being integrated through artificial intelligence.

First Scene of Intelligence

The future will be the world of "intelligence," and everything in the world will be connected and interact with each other, resulting in a lifestyle that moves at your will.

When you are still asleep, soft music will slowly play, the bedroom curtains will automatically open, and the warm sunlight will lightly pour into the room, awaking you to start a new day.

When you get up and wash up, the nutritious breakfast will be ready, and after the meal, the audio turns off automatically to remind you that it's time to go to work.

When you come home from work, the garage door automatically opens by the time your car arrives. When you get home, the smart home is already working — the light is on the moment you open the door, the most suitable room temperature is ready, and the bath water is just right ...

It is artificial intelligence that provides these thoughtful applications.

Intelligent sweeping robot walks around every corner of the room, takes away dust, shredded paper and clothing fibers, and makes the room clean and tidy. Different from the general sweeping robot, the intelligent sweeping robot can automatically identify the real situation of the room, automatically generate the most efficient route, avoid the obstacles and complete the cleaning all by artificial intelligence.

Let's change the scene to the street. Many years ago, there were already cameras on the street. However, faced with emergencies such as traffic accidents, road surface water and electric light failures, most of the time people rely on active response and passive maintenance.

With the application of artificial intelligence, the cameras, street lights and other things all over the street will be connected into a whole. The street lights will be turned on or off according to light conditions, the road surface water will be automatically monitored and the system will immediately notify the relevant departments to deal with it, and the vehicle diversion in real time will be guided during road congestion.

Artificial intelligence is everywhere, being in the first scene of all changes.

Innovation Through Artificial Intelligence

Mankind has experienced the steam engine era, the electrification era, and the information era. Now we have come to the intelligent era which is guided by the Internet, the IoT, big data, cloud computing, machine learning, and the knowledge revolution.

The current popular technologies such as IoT, 5G and big data have motivated their greatest potential under the integration of artificial intelligence.

Behind all the changes, there is the shadow of artificial intelligence.

The industry has a consensus that artificial intelligence infiltrates into all application scenarios and penetrates everything in life.

Each year's Smart China Expo brings together the world's innovations, which, in one way or another, are all linked to artificial intelligence.

In 2020, in the shadow of the COVID-19 pandemic, smart medical became a popular segment. New applications and medical technologies such as teleconsultation, Internet hospitals and AR telesurgery have emerged in the market.

"Artificial Intelligence + Medical" plays a linkage role in various aspects, such as smart consultation, smart triage, pharmaceutical R&D and precision medicine. Smart diagnosis can assist medical decision-making, and smart diagnostic equipment can advance precision medicine. Further, AI can also break through existing bottlenecks in biopharmaceuticals, targeted therapies and other applications.

Machine intelligence has been implemented in the health care industry, from electronic medical to mobile medical, and gradually transformed into the current smart medical. In the future, it will develop towards intelligent medical care with cognitive capabilities.

Smart healthcare under the Internet of Everything can automatically collect real-time data, sort and analyze various environmental information that it comes into contact with.

At the same time, artificial intelligence has become an integral part of the Industrial Internet.

In the intelligent workshop of Chongqing Chang'an Automobile, the mechanical arm is constantly waving on the empty assembly line, and the whole production elements of humans, machines, materials, methods and environment are connected through the fully connected factory platform to achieve unmanned production by using artificial intelligence and intelligent control, to carry out real-time management of personnel, equipment, production capacity, energy consumption, logistics and other information, to achieve unmanned and transparent factory management.

In this new era, people have fundamentally changed the way that ask questions about AI. Previously, it was customary to ask, "What can AI change?" Now, people are more interested in knowing, "What else won't be changed by AI?"

Artificial intelligence is innovating everything.

The Future of Artificial Intelligence

Led by artificial intelligence, the technology is rapidly iterating.

Technologies such as 3D sensing cameras, 5G, artificial intelligence cloud services, AR cloud, augmented intelligence, autonomous driving, biochips and decentralized networks have all come out of the emerging technology Hype Cycle to peak at high expectations.

While other technologies are being iteratively upgraded, AI technology itself is also being upgraded on all fronts.

In the 2020 Gartner Hype Cycle, AI remains the absolute protagonist, and the coverage of AI potential continues to grow, such as composite AI, generative AI, responsible AI, AI-enhanced development, embedded AI and enhanced AI design.

Among other things, formative AI senses and dynamically responds to changing situations, providing real-time interactive feedbacks to visual design/usability design (UI/UX) designers to improve the usability of software and intelligent products. Gartner predicts that formative AI will be used to simplify and fine-tune the creation of mathematical and machine learning models over time.

AI-enhanced design has the potential to transform the way digital and smart connected products are designed, produced and sold. Composite AI brings together or combines different AI technologies to improve the accuracy and efficiency of learning. Embedded AI has the potential to improve the accuracy, insight and intelligence of current and next-generation sensors. Generative AI is the technique most commonly used to create "deepfake" video and digital contents. Responsible AI helps organizations make more ethical and balanced business decisions by reducing deviations.

From voice-driven personal assistants like Siri and Alexa to self-driving autonomous vehicles, AI has been spawning new offshoots of technology based on market needs. Many technology giants such as Apple, Google, Huawei, Baidu and Xiaomi have been investing on the long-term growth potential of artificial intelligence.

In the course of these applications, the new directions derived are not only to tap the potential of the current technology, but also to modify the path of the intelligent era.

As artificial intelligence continues to iterate and upgrade, new technologies, new applications and new life scenarios will occur to respond to the call of the times.

Chapter 2

Trends: Key Navigation in the Intelligent Industry

For many traditional industries, this is the last stop for change; for many new industries, this is a brand new starting point for boldness and strength. Whether it is the end or the starting point, artificial intelligence has given the opportunity to redefine. New industries can soar along with the trends, traditional industries can regain a new life, every inch of the air in the intelligent era is filled with the echo of the innovators.

For the majority of people in the whole era, every call is reverberated, but the direction is confused. It is time to find out a key navigation for the intelligent industry of this era.

Section I

Jump-Start 5G, Key Layouts Affecting the Global Intelligent Industry

At the last stage of science lies the imagination.

— Victor Hugo

With the official issuance of 4G license by the Ministry of Industry and Information Technology (MIIT), China officially entered the 4G era at the same time as the rest of the world did.

The biggest expectation people had for 4G at the time was that the network speed might be a little faster. Eight years later, we realize that 4G has changed the entire world far beyond speed. As all aspects of life and work, including communication, consumption, travel, finance and entertainment, have been profoundly affected and changed by 4G.

How exactly will 5G change the world? Perhaps everyone has their own answers, while inevitably possessing their own one-sidedness and limitations in front of the times.

Although the understanding varies, it does not affect 5G, in any way, to be the most critical deployment for the global intelligent industry.

According to the 2020 Gartner Hype Cycle, 5G technology which is maturing has been removed from the emerging technology hype cycle and successfully transitioned to the next phase.

Advance Exploration of "5G+"

In people's conception, all industries, together with 5G, can complete a qualitative change and achieve an industrial leap.

With this conception, 5G unmanned mines, 5G intelligent transportation, 5G intelligent medical care, 5G ultra-high voltage substations and 5G drones have become the new runways towards 5G.

Although there are still 2–5 years for 5G technology to enter maturity on the Gartner Hype Cycle, top companies around the world have been running wildly on the track and conducting explorations in various fields in order to seize the pulse of the smart era in advance.

Chapter 2 Trends: Key Navigation in the Intelligent Industry

At the 2020 Smart China Expo Online, Huawei exhibited the Kunpeng computing industry ecosystem that integrates software and hardware. With "5G + Kunpeng + Ascend + Cloud" as the core, the ecosystem runs through the entire chain of IT infrastructure to upper-level application systems. Good results have been achieved in terms of increasing efficiency and reducing resource usage costs in the fields of manufacturing, medical treatment, electricity and transportation.

Meanwhile, China Mobile Chongqing released the "5G + Industrial Internet Platform," a 5G+ Industrial Internet Laboratory jointly established by the leading local manufacturing enterprises, to build 5G factories by deeply integrating 5G, big data, cloud computing, IoT, artificial intelligence and other emerging technologies. In the fields of manufacturing, logistics supply, equipment after-sales and other scenarios in the automotive, equipment, energy, food, and pharmaceutical industry, China Mobile Chongqing will provide integrated solutions for the application of 5G technology to the industrial manufacturing industry, and to explore the integrated and innovative applications of 5G in the development of enterprises' smart manufacturing, remote control and smart industrial park.

In the Smart Shared Logistics Center of the Airport Industrial Park of Cuntan Bonded Port, "5G + Smart Logistics" is playing a full role. AGV intelligent carts are transporting goods, bins are moving automatically on the conveying line, and stacker cranes are orderly stocking and picking up goods. Outside the warehouse, self-driving cars are transporting the goods to the nearby factory. At present, the Smart Shared Logistics Center has built an Automatic & Storage Retrieval System with 38,000 storage capacity, electronic material warehousing, distribution material identification systems and unmanned intelligent transportation systems.

If you are tired of traditional offline shopping, AR interaction, cloud VR shopping with coupons, VR shopping guide and 5G live streaming will take you to experience a new online shopping. Chongqing Wanxiangcheng has joined hands with China Telecom Chongqing to take the lead in landing the "5G + MEC" business cloud platform in the southwest China. The platform, based on the high bandwidth and low latency of the 5G network, allows consumers to experience rich AR and VR virtual scenes and participate in entertainment interactions, so they can enjoy an immersive shopping experience at home.

In addition, Qualcomm shared the world's first robot platform that supports both 5G and AI robot platform, namely the Qualcomm Robotics RB5 Platform. It is Qualcomm's highly integrated overall solution designed specifically for robotics. The platform provides a combination of hardware, software and development tools that enable developers and vendors to build next-generation robots and drones with high computing power and low power consumption to meet the requirements of consumers, enterprises, protection, industrial and professional services sectors.

In the UK pavilion at the 2020 Smart China Expo online, the Wireless Logic Group showcased the IoT Global Connected Platform. On the platform, devices connect everything from electronic point-of-sale terminal systems in retail stores, to car behavior monitoring, and then to off-grid solar installations in remote rural areas of East Africa. In every 18 seconds, a new device, new terminal or new asset is connected to the platform, and with the addition of 5G, connectivity efficiency will be even higher.

At present, the new technologies represented by 5G open the era of Internet of Everything, and new industries, new businesses forms and new models are accelerating the penetration and integration with various industries.

Different Focuses of Major Countries on Competition

The year 2020 was an extremely important year for world history. Under the impact of the COVID-19 pandemic, the world has accelerated the pace towards the intelligent era.

At present, a new generation of network information technology continues to innovate and breakthrough. Digitalization, networking, intelligent development, and digital transformation of the world economy also continue to accelerate.

In the new round of sci-tech revolution, 5G technology has become the catalyst for change in the intelligent era, and the breakthroughs of core AI technology and commercialization applications will be the focuses. The road to commercialization of artificial intelligence will rely on 5G for faster and stronger support.

In terms of the formulation of 5G standards, the layout of 5G base stations, and the exploration of 5G applications, all countries are accelerating their deployment and have formed three major markets in North America, Northeast Asia and Western Europe respectively. In terms of the key factors, such as spectrum availability, 5G deployment progress, government-related policies and financial support, investment of industries and enterprises, and market space, we could safely assume that China and the U.S. are the first echelon, followed by Japan, South Korea and Europe.

At present, China has raised 5G to the national strategic level. With the leading of the government and efforts of enterprises in tackling down tough problems, the comprehensive advancements of R&D, network construction and industrialization have been realized. China has opened a new infrastructure to develop vigorously in 5G networks, artificial intelligence, Industrial Internet, Internet of Things, data centers and other fields so as to build the cornerstone of the new economy, provide strong support for the development of the digital economy, guide the deep integration of the digital economy and the real economy, promote high-quality economic development, and accelerate the construction of digital China.

Similarly, the U.S. government is actively promoting the development of planning 5G technology, and the U.S. takes the lead in 5G millimeter-wave research. However, due to the blocked direction of millimeter-wave development and the lack of abundant spectrum resources, mmWave 5G availability is still being explored in the direction.

On April 3, 2019, South Korea's three telecom operators, SK Telecom, Korea Telecom KT and LG U+, were the first to announce the launch of 5G services, turning

South Korea into the world's first country to implement commercial 5G networks. South Korea attaches great importance to 5G, however, due to the limitation of its population, the market is small and insufficient to cultivate a wide range of 5G application scenarios.

Japan's 5G industry is developing at a slightly slower pace, ranking low among major global powers, and its three major telecom companies are relatively conservative about building commercial 5G networks. However, Japanese communication equipment manufacturers are more enthusiastic about 6G research and development.

In the EU region, the overall development of 5G construction is slow with a narrow range of applications and uneven development across EU member states.

With the characteristics of real-time online, interconnection of everything, high concurrency, low latency, high reliability, high broadband and high frequency, 5G will bring revolutionary changes in the industrial field, overturn traditional production methods, further expand the field and space for the development of digital economy, and bring new development opportunities for the intelligent manufacturing industry.

It is a global consensus that "4G changes life and 5G will change society." 5G will not only bring changes in speed, but also serve as the underlying foundation for the Internet of Everything, which can be superimposed on various applications and innovations, deeply integrated into smart life, smart manufacturing, smart services and other scenarios. 5G will become a new engine for global economic growth and trigger a series of industrial revolutions.

5G Enhances the Intelligent Industry

Compared with previous generations of communication technology, the biggest change 5G brings to human society is the extension from "Human to Human" information exchange to "Thing to Thing" connection and interaction. Once everything is connected, the whole world will have a wonderful chemical reaction.

As 5G continues to spread, everything will be connected to the cloud and interact with each other. We are moving into a new era of intelligent cloud connectivity powered by 5G and AI.

At the 2020 Smart China Expo Online, the world's most cutting-edge technology, the most pioneering application exploration and the most visionary future predictions, in a few strokes, outlined the countless possibilities of the intelligent era.

From 5G factory to Industrial Internet, from silicon-based optoelectronics to 8K new display, from smart governance to smart city, from factory farming to smart agriculture, all industries have found their positions in it.

The freezing of R16 standard and the promotion of R17 provide better support for the characteristics of 5G wide connectivity. Through 5G we can access to more devices, AI can control more devices, and the application scenarios of AI are also broader.

5G empowers the intelligent industry to enter a high-speed development phase, and thrives in all walks of life, by carrying out practices on different paths. Technology elites are expressing their views and discussing the call of the new era.

The call of the new era corresponds to absolute commercial potential. IDC estimates that the overall size of China's AI market was about $6.3 billion in 2020 and will be $17.2 billion in 2024.

In the field of algorithm theory and platform system development, Europe and the U.S. are still at the forefront of the world and exploring more cutting-edge technologies, such as core algorithms, key equipment, high-end chips, major products and systems.

Machine learning algorithm is the hot spot of AI. Open-source deep learning platform is the core driving force of AI application technology development, and open-source framework has become the focus of the layout of international technology giants and unicorns. At present, open-source frameworks that are widely used internationally include Google's TensorFlow and Microsoft's DMTK.

China has adopted a different path and achieved breakthroughs in core technologies such as voice recognition, visual recognition and information processing, which has a broad application market environment. Gartner's report on 2020 Hype Cycle for ICT in China added a number of new technologies and new business forms led by China and risen in China, including edge computing, workflow collaboration, live e-commerce, data middle platform, middle platform architecture, cloud security technology, and blockchain technology.

Both the self-directed strategies of global enterprises and countries' overall strategies show that the entire human society has reached a consensus on the arrival of the intelligent industry era. Countries and enterprises vary in the focuses of deploying the intelligent industry, but in 2020, they all made some key breakthroughs.

Riding on the global wave of intelligence, 5G technology brings new opportunities to countries and enterprises. Although there is no unified answer to the ultimate imagination of 5G, one thing is certain. That is, the whole world is rushing forward with a common goal.

Section II

Foresee the Next 30 Years of Intelligent Industry from New Energy and Carbon Neutrality

> True life and true truth override opposing concepts, such as money and faith, mechanics and mind, reason and piety.
>
> — Herman Hesse

Is the intelligent era still an imaginary future?
Nope. The dream has turned into reality.

The beginning of a new era is always filled with opposing concepts: virtuality versus reality, enthusiasm versus conservatism, indulgence versus restraint, openness versus taboo, and exploitation versus protection.

We need to develop the digital economy, but also need to prevent detachment from reality. We need to explore autonomous driving, but also need to consider traffic safety. We need to decipher the genetic code, but also need to worry about the ethical crisis. We need to accelerate energy development, but also need to pay attention to environmental protection.

However, technology is born to resolve confrontations and conflicts.

Rapid development, emission reduction, resource development and environmental protection are superficially full of contradictions between advance and retreat, taking and giving, but essentially they are deep-seated changes in development models and methods that need to be innovated.

The COVID-19 Pandemic Brings New Inspirations on Green Development

Throughout 2020, the world was under the impact of the COVID-19 pandemic. As of December 31, 2020, the total number of confirmed cases in the world has reached 81,475,053 with death toll rising to 1,798,050.[1]

---

[1] China News, *WHO Reports That the Official Number of Covid Deaths Was 1.8 Million*, January 1, 2021.

And the epidemic is mutating and spreading in different countries and regions, with no signs of abatement going into 2021.

The COVID-19 pandemic will leave a lasting mark on the world economy, bringing permanent change and teaching human beings an important lesson.

In the future, there may be a greater ecological crisis lurking. In order to win the initiative in advance, it is imperative to change the development mindsets and adhere to the concept of green development.

In 2006, the *New Oxford American Dictionary* named "carbon neutrality" its "Word of the Year," which is a testament to how a growing environmental protection culture is "greening" human language.

Carbon neutrality, or net zero emissions, means that the carbon emissions necessary for human economic and social activities are captured and utilized or sequestered through forest carbon sinks and other artificial technologies or engineering means, so that the net increase in greenhouse gas emissions to the atmosphere is zero.

On September 22, 2020, in the 75th UN General Assembly, President Xi Jinping proposed that China would strive to achieve peaking carbon dioxide emissions before 2030 and carbon neutrality before 2060. This is not only China's national policy to actively address climate change, but also a national strategy based on scientific evidence, both as an action goal from reality and as a visionary long-term development strategy.

The progress and innovation of energy technology is the fundamental driving force of energy revolution and transformation development, and is also the key driver and inevitable choice to achieve the goal of "carbon neutrality."

As early as 2009, China released the *New Energy Industry Revitalization and Development Plan*, both for the development and utilization of renewable energy sources such as solar and wind power, as well as for the transformation of traditional energy systems such as coal chemicals, with an expected investment of over 3 trillion yuan by 2020.[1]

From new energy to carbon neutrality, it is not only an upgrade of concept in the green development, but also a direction for developing the intelligent industry in the next three decades.

At the 2020 Smart China Expo Online, Liu Zhenmin, UN Under-Secretary-General expressed, "Developing smart technologies can help build a more inclusive and sustainable future. Making good use of smart technologies can help eradicate hunger and poverty, promote sustainable agriculture, expand access to education, improve public health, build smart cities and sustainable infrastructure, and optimize public services. Smart technologies will continue to empower the *2030 Agenda for Sustainable*

---

[1] People's Daily Online, *New Energy Industry Revitalization and Development Plan Will Be Introduced*, May 21, 2009.

Development and help achieve the UN Sustainable Development Goals." [1]

At present, China is vigorously promoting the transformation, and upgrading the high-quality development of domestic industries. The upgrading of industrial structure can reduce carbon emissions and improve carbon emission performance, while carbon emission policies have a driving effect on the upgrading of industrial structure.

A new round of investment in China's new energy industry has been launched, and is included in the 14th Five-Year Plan of many cities. It is expected to add more than 1.2 billion kilowatts of solar and wind power in the next 10 years, which will bring a huge market of 12 trillion yuan.[2]

Era of "New Energy + Intelligence" Starts

People often sigh when the times abandon us without even saying farewell. In fact, it is not true. When the dawn is coming, the watchmen are never absent. The sound of banging clappers is the sound of the footsteps of the coming new era.

From new energy to carbon neutrality, every step in between is traceable.

At present, the global energy revolution has entered the multiplier stage, and the era of "new energy + intelligence" has started. Technology companies around the world are thinking about two questions: How can the enterprise itself achieve 100% renewable energy utilization? How to use its own cutting-edge technology to drive the whole society to accelerate carbon neutrality?

In the past decade, several international technology companies such as Google and Apple have created several successful stories by leading the trend of 100% renewable energy.

At the 2020 Smart China Expo Online, there were many "watchmen" of the new era, as well as many new technologies and products of green development around the world. They are like the spring birds that bring the happy news of blossoms, conveying the message of the new era to everyone.

In 2021, China has officially entered the era of "grid parity" for wind power and photovoltaic, which makes it possible for Internet enterprises to adopt 100% renewable energy. At present, Alibaba, Qinhuai Data, Wanguo Data and Baidu have already achieved large-scale market-based green power trading in some of their data centers.

Energy conservation starts with the data center. In Tencent's Marina Building and data center, artificial intelligence and cloud computing are applied to carbon emission reduction, and the T-Block energy-saving technology has been iterated to version 4.0.

Take Marina Building as an example. The ecological ceramic permeable tiles paved on the 8,000-square-meter square can absorb and purify a large amount of

---

[1] Yang Ye, ShangyouNews, *Wisdom at SCE|Liu Zhengmin: Intelligent Technologies Empower 2030 Agenda for Sustainable Development,* September 15, 2020.

[2] Zhu Yanran, YiMagazine, *The New Round of Investment in the New Energy Industry Will Reach 12 Trillion Yuan with the Consumption as the Key,* January 24, 2021.

rainwater, which can be used to water the flowers and plants in the building. The ceramic granule layer on the roof of the north and south towers can achieve the effects of purifying rainwater, slowing down the flow rate of rainwater and reducing the flood peak. The intelligent lighting system installed in the office area can save about 1,236,100 kilowatt hours of electricity per year.

Tencent's Tianjin Data Center is developing a scheme for energy-saving applications based on the principle of waste heat recovery. According to this scheme, if all the waste heat of the center in winter is recovered, a 460,000-square-meter area or more than 5,100 households can benefit from it. The amount of carbon dioxide emission can be reduced for 52,400 tons, which is equivalent to planting 2,864,000 trees.

A pleasant building should not only be safe, sturdy, comfortable, convenient, green and energy-saving, but also have a certain level of disaster resistance.

At the 2020 Smart China Expo Online, Arup, a British company, has planned a more livable and smarter city by integrating data and information-based models of buildings. It avoids material waste and improves design and construction quality through BIM modeling technology and collaborative adoption of 3D models. Arup has designed and built many smart buildings using BIM modeling technology, including the Beijing Olympic Stadium, Guangzhou Tower and Raffles City Chongqing. In addition, Arup has implemented a "sponge city" strategy for 3.5 square kilometers of land in the southern part of Beijing Shougang, making the new integrated service area of Shougang high-end industries the first C40 climate-positive development in China which is resilient to natural disasters caused by rainwater.

At present, the heat wave of consumer Internet hasn't faded and the picture scroll of the Industrial Internet of energy industry is also slowly unfolding.

Like other industries, the energy industry is moving from unilateralization to marketization and from primitiveness to digitalization. In the process of transforming and upgrading the energy industry to intelligentization and digitalization, technology companies are using cloud computing, big data, IoT, 5G, artificial intelligence and other high technologies to empower the energy industry and facilitate carbon neutrality.

Among them, Tencent Cloud launched four new products in the field of intelligent energy, kicking off the prologue of technology enterprises' driving social carbon neutrality. Comprehensive energy open platform, energy cognitive brain, power portrait of enterprise users and smart gas station will provide diversified solutions for the energy industry and promote digital transformation of energy enterprises. In addition, Baidu, Ali, Huawei and other leading companies also launched a variety of energy Internet-related products.

Green Future of the Intelligent Industry

As we all know, if you want to sustain a long-term development, you can't just focus on the present benefits.

No bright imagination about the future of humanity can be built on a scarred planet.

In these years, the world has promoted the deep development of intelligent industry, turning the imagination in science fiction works into a touchable reality. However, where will the intelligent industry go in the next 30 years?

At the 2020 Smart China Expo Online, Robin Li, founder, chairman and CEO of Baidu Group, stated, "The future of industrial intelligence will eliminate traffic congestion, improve the efficiency of production and work, reduce waste of resources, realize intelligent and convenient services, and build a more civilized and safer intelligent society."

"Intelligence" is the ground color of green development, smart homes, smart factories, smart cars and smart cities. All intelligent imaginations are for the green and high efficiency of the whole world.

In the next three decades, with more advanced technological innovations, deeper industrial synergy and richer data resources in the intelligent industry, AI will play a huge role in many fields, such as energy, weather, environment and water resources.

In terms of the energy field, the intelligent technology will not only drive the energy industry to digitalize, but will also play an irreplaceable role in the development and application of new energy sources, giving rise to more environmentally friendly energy sources.

In terms of meteorology, at the 2018 Smart China Expo, the four systems based on the deep development of artificial intelligence, developed by Chongqing Meteorological Bureau: "Tianshu" "Tianzhi" "Zhitian" and "Yutian" had made the debut. With the further development of cutting-edge AI technology, these systems are being upgraded to be effective in disaster prevention and mitigation in daily life and production.

In terms of water resources, China's first full data integrated water service platform has entered the fast track. Ali Cloud and Chongqing Water Affairs Group announced that they would jointly develop Chongqing's intelligent water services to achieve the deep integration of new technologies such as big data, cloud computing, IoT, artificial intelligence and the water service industry to improve processing efficiency and save water resources, so that the water service system would become more scientific, smarter and greener.

In the future, the application of cutting-edge technologies, represented by artificial intelligence, will become popular in energy saving and emission reduction in industries. The potential in addressing the major challenges of the Earth will be further motivated.

"No green, no development," will become a new standard in the new era. The real green development is inseparable from the support of the smart industry.

Section III
Industry Without Boundaries, Artificial Intelligence Throughout All Scenarios

The boundaries of science are like the horizon, the closer you get to it, the farther it moves.

— Bertolt Brecht

In the past, there was a clear distinction between one industry and another. Everyone continued to study and innovate independently in different fields. They knew each other but seldom interacted with each other.

After more than 20 years of repeated transformations by the Internet wave, every field is changing, industries are changing, enterprises are changing, and our lives are changing. Different industrial elements break through the walls and the cross-border integration of different industrial fields has instead become a brand-new innovation methodology.

Into the era of intelligent industry, this cross-border integration continues to accelerate. Artificial intelligence is playing the role of ubiquitous fusion agent who is erasing the last boundaries and barriers between industries, so that all industries have a new look with the common blood flowing through them.

Artificial Intelligence Is Everywhere

When describing the future of any company focusing on AI innovation, they will build a smart life scenario with perfect experience based on their imagination. However, in the era of rapid innovation and change in the intelligent industry, a single technology product always enters consumers' daily lives quickly, but the "perfect scenario" never comes to the public. After all, without the ability of large-scale technology assembly and deep-level technology linkage, enterprises and products only constitute a part of intelligent life .

Chongqing has successfully held the Smart China Expo for three consecutive years while continuing to bring together the world's cutting-edge technologies to create a wonderful future world.

In Lijia Smart Park of Chongqing Liangjiang New Area, black technologies are gathered and smart life is within reach. Artificial intelligence is ubiquitous, running through all the scenes of people's lives, such as food, clothing, housing, transportation and entertainment.

At present, Lijia Smart Park has "one park and five zones," including Lingjiang ACGN (animation, comic, game and novel), Cloud Forest, Geek Community, Lakeside Smart Core and Innovation Center, to provide a one-stop experience of smart life in the future city.

When you are thirsty, the robot will make a cup of coffee for you. When you are hungry, the robot will cook a bowl of authentic noodles for you. When you are sleepy, you can take a short rest on the customized smart bed with various health management functions. When you want to sing a song, the AR technology will "create" a whole band for you and you will be the lead singer. When you want to see the scenery of Chongqing, the 5G real-time image will guide you "one-stop" tour of Chongqing's popular attractions. If you want to exercise, VR games take you to play hockey, play soccer ...

In addition to these scenes of life that are novel to us, AI is also present in many fields of agriculture, industry and services.

Farmers used to have dark faces and rough hands. Agriculture used to be the stereotype of hardworking from the sunrise to the sunset in the field. Agricultural production and artificial intelligence represent the source and future of human civilization respectively. In addition to the simple application of technology, it were difficult to have a deep integration between agriculture and artificial intelligence. In fact, artificial intelligence has already penetrated into all aspects of modern agriculture.

In the exploration of smart agriculture, the farmland is surrounded by sensors. Key information, such as temperature, humidity, soil nutrients, and crop growth in the intelligent monitoring system, will be converged into a strip of data, thus driving commands. According to the commands, weeding robots shuttle through the farmland, drones spray pesticides and fertilizing robots apply fertilizer according to the nutrient level of different plots. When the crops are ripe, the harvesting robots automatically start working, and the farmer can learn everything that happens in the field just by holding his cell phone.

Besides the common robots in sci-fi movies, AI technology penetrates the Industrial Internet, Created in China, smart robotic arms and 5G factories, driving the integration of informatization and industrialization, and promoting the application of the ultra-densely connected Internet of Things, Internet of Vehicles, Industrial Internet and smart manufacturing.

In the fields of public health and professional medical care, the application of AI technology is also becoming increasingly deep. In the prevention and control of the COVID-19 epidemic and treatment, artificial intelligence has provided solutions for the construction of smart hospitals, and provided ideas and technical supports for the construction of smart public health. Various places and medical institutions at all levels have carried out the construction of smart hospitals and Internet hospitals, applying online appointment, telemedicine and other important initiatives to improve medical services. The concepts of Internet hospitals, artificial heart, remote surgery have become buzzwords, and smart medical is just around the corner.

Artificial intelligence is everywhere. It has landed in the application scenarios of all walks of life such as unattended service desk, autonomous driving, intelligent access control and cloud viewing, which are all the credit of artificial intelligence.

Industrial Integration Becomes the Inevitable Path

In 1765, Hargreaves invented the Jenny spinning machine, which opened the industrial era. Since then, various new technologies and inventions emerged and were rapidly applied to many fields, such as industrial production, the popularity of electric energy, the emergence of the internal combustion engine and the invention of modern communication... Many revolutionary innovations have led the inhabitants of this planet into a new modern society.

Within industry, science and technology are being organically integrated. People are acquiring new technologies through scientific research, and in turn are facilitating the application of theory. This process of change has continued and has been doing addition. The emergence of artificial intelligence is making the addition into multiplication, making the penetration more thorough and the change more disruptive.

Artificial intelligence penetrates all industrial scenes, and accelerates the penetration and integration of high-tech and the extension and integration of industrial innovation. Industries, such as agriculture, industry, service, information and knowledge, penetrate each other, contain each other and integrate development in the same domain, industry chain and industry network. The integration of the tangible with the invisible, the dominance of the high-end over the low-end, the upgrade from the backward to the advanced and the development from the vertical to the horizontal, will facilitate the high-end industries to become the components of the high-end industries and realize industrial upgrading.

The independent state of each industry was broken, forming a new trend and new form of industrial integration. The reasons are mainly because of the increase in electronic data, the popularity of mobile interfaces and the development of artificial intelligence.

At present, the interactive integration of the service industry and manufacturing industry has become the main way of industrial integration, such as the integration of industrial design and manufacturing. However, in order to achieve continuous upgrading of products, we have to work on industrial design, and creativity, technology, data and artificial intelligence are indispensable elements.

At the 2020 Smart China Expo Online, top experts, scholars and enterprise representatives in the field of Industrial Internet gathered to explore the new paths of Industrial Internet technology innovation and ecological construction, and to discuss how to accelerate the integration of Industrial Internet with the real economy.

Chapter 2 Trends:
Key Navigation in the Intelligent Industry

On the eve of the opening of the 2020 Smart China Expo Online, the 5G + Industrial Internet Lab, established by China Mobile Communications Group Chongqing Ltd. in conjunction with a number of leading local manufacturing enterprises, was officially inaugurated at the 2020 Industrial Internet Innovation and Development Conference.

The lab will focus on actual application scenarios such as production workshops of manufacturing enterprises, provide integrated solutions for 5G technology application in industrial manufacturing industry, and on this basis, further explore the integration and innovative applications of 5G in the development of smart manufacturing, remote control and construction of smart industrial parks, and continuously promote the digital transformation and upgrading of traditional industries.

In recent years, the development of China's rural industrial integration has been accelerated, and the trend of developing multiple forms of "agriculture+" is gradually taking shape. The symbiosis of ducks and rice, central kitchens, leisure agriculture and smart agriculture have emerged in large numbers. The applications of data in decision-making, management and innovation, constantly strengthen data resources to the "gather and share" level.

At the 2020 Smart China Expo Online, Chongqing Municipal Agriculture Commission launched a big data platform for agriculture and rural areas, focusing on the agricultural products quality and safety traceability and agricultural inputs supervision platform. The platform realizes intelligent monitoring and online certification, and promotes the integration of production, products and services of agriculture through intelligent agricultural production, business networking, management data and online services.

Artificial Intelligence Shapes the New Industrial System

Without barriers, everyone could deeply feels the changes in the world standing on the waves of the times.

New business forms, new trends and new concepts are constantly pushing forward the boundaries of human imagination.

Experiments have shown that quantum entanglement, which Einstein called "spooky action at a distance," occurs between particles as far as 1,200 kilometers away and at speeds faster than the speed of light, suggesting that the connection of everything may be the essence of the universe.

This is also the essence of industrial integration. Industries, such as service industries and manufacturing industries, emerging industries and traditional industries, virtual economy and real economy, software development and hardware production, link with each other, undertake new trends and transfers of industries, and give birth to new business models, forming a new pattern of multi-form, diversified, multi-channel and multi-level industrial integration development.

The new era is undergoing profound changes, and the penetration and integration between industries clearly show the development vision for the intelligent era. In different industrial fields, industrial integration evolves in different ways, and through the grafting and sublimation of science and technology, industries are interconnected and integrated, and the boundaries are gradually blurred, which cannot be matched in the traditional industrial classification system, so that new industries will evolve. Eventually, it will lead to a new industrial system and move towards the integration of industries without boundaries.

No matter how the business forms and trends change, one thing is certain that the new industrial system will certainly rely heavily on information technology, especially data and artificial intelligence.

Borderless industries are directly driven by consumption and require a dynamic combination of resource elements, industrial synergies and operational approaches for changing demands. The diversification and changes of consumer demands have caused these new industries to be in a state of uncertainty at all times, and the frequency of renewal is accelerated. In order to meet these needs, the development of interdisciplinary sciences has become an inevitable trend.

The capability of artificial intelligence to penetrate all scenarios will provide a breeding ground for the cultivation of interdisciplinary sciences and will also accelerate the borderless integration of industries.

Accordingly, the "laboratory economy" will also take the advantage of this trend, with enterprises as the leading factor and laboratories as the carrier, forming a market-oriented technological innovation model. From R&D investment to core technology and then to industrial advantages, this new industrial model is becoming an important force driving the vigorous development of the intelligent industry.

In the future, the borderless industrial ecosystem will be indexed by artificial intelligence, gradually linking various sectors of each industry and forming a new industry chain and a new system of network-type extension. Multiple partners in the ecosystem rely on each other, coexist and prosper, and produce synergistic effects to achieve co-evolution.

Further in future, perhaps the only thing the industries can be sure of is that they are in a constant process of change and will have new names at any moment.

Section IV

Digital Life and Artificial Intelligence Redefine Social Relationships

Numerous personnel changes in the time of the embryo.

— William Shakespeare

People living in this age may not realize that our ancestors of thousands of years ago, year after year, generation after generation, were full of endless repetitions and lack of changes. They took over their fathers' hoes and plowed their ancestors' land. They had never experienced as many changes in a lifetime as we experience in a year.

The era we live in, every today that is happening, is different from any day in the past, and everyone in it can always perceive the speed of change.

The development of science and technology has become the core force that dominates the changes in the world, as if there were a pair of invisible pens in the vast universe, painting people's every dream and wish into a new reality.

Against the backdrop of the 2020 COVID-19 epidemic lockdown, we enjoyed the springtime and viewed flowers through Cloud VR. We met friends who were thousands of miles away through a video call. We enjoyed many services, such as medical treatment, shopping and grocery shopping at home.

In these digital life scenarios, artificial intelligence is reshaping our lives.

Ubiquitous Digital Life

Once upon a time, carriages and horses were slow and letters travelled far away, only a handful of friends could be connected in a lifetime, and many partings were destined to last forever.

Now, high-speed rail moves faster, video call brings us closer, friends interact every day, and the ends of the earth seem close at hand.

There is no era like today, where people can have two worlds at the same time, an offline real world and an online virtual network world. Both of them merge with each other and exist in harmony, constituting people's digital life.

In this era, smartphones have replaced wallets and keys as essential things. Communication, transportation, shopping, reading and entertainment have long been integrated with the smartphone. The smartphone has become a carrier of digital life.

With the integration of the digital and physical worlds, production and life are being digitally reshaped with online to offline penetration and compatibility. With the help of artificial intelligence technology, different digital sectors are linked together to constitute people's digital life. "In the future, pure 'offline life' and 'traditional industries' will not exist, " according to Ren Yuxin, Chief Operating Officer of Tencent.[1]

In the digital life, the way people communicate with the world has changed dramatically, overturning the traditional life, and the online and offline life tend to be integrated.

Tencent's Photon Studio implanted the Miao characteristics of Chi You Jiu Li City in Pengshui county, Chongqing into games, such as "Peace Elite" and Tencent's chess products. Offline scenic spots also combined the game IP to create special scenes to enhance visitors' experience, such as "Happy Teahouse."

The combination of "game + cultural tourism" presents a new experience. Through the co-branded cultural and creative products, young people fall in love with the traditional culture of Miao embroidery, Miao songs and batik. There are experiential publicity for hundreds of millions of people every day.

The community in reality already has a smart brain.

In the "Smart Community" Pavilion, a smart community of Chongqing Lijia Smart Park, the camera is equipped with a smart detection system for objects thrown from high buildings, whose timely location tracing and dynamic capture of objects falling from high buildings help to organize the chain of evidence and quickly find the perpetrators when the incident of objects thrown from high buildings injures people. The smart access control automatically identifies the identity of residents and the wearing of masks, and only when the conditions are met will they be allowed to pass. The fall detection system automatically identifies the human body postures. Once the elderly falls at home, community managers and relatives of the elderly immediately receive the alarm signal and take first aids... All the smart communities with these smart application scenarios come from the 42 smart communities that have been built in Liangjiang New Area.[2]

For the residents of Liangjiang Xinchen Yunding District, the intelligent activities are not confined to the demonstration in the laboratory, but become a daily routine. People can open the door or their houses through facial recognition, share pass code on smartphones, or quickly contact the property management staff by smart Apps. Some smart Apps can also monitor temperature and soil humidity to water plants automatically.

On the wooden walkway in Lijia Smart Park, a mechanical arm called "Taiji Push Hands" is pushing back and forth, just like a Taiji master. People can compete with it to

---

[1]Sun Lei, Shangyou News-Chongqing Business Daily, *Tencent's Ren Yuxin: Pure "Offline Life" and "Traditional Industry" Will Not Exist*, September 15, 2020.

[2]Wang Qian & Chen Xiang, Chongqing Morning Post, *42 Smart Communities Built in Liangjiang New Area Allowing People to Enjoy Life with a Cell Phone*, July 22, 2020.

correct the Taiji postures and achieve the effects of strengthening the body.

Our lives are given a new look with the support of artificial intelligence, but Taiji is still Taiji and scenic spots are still scenic spots. It is just the technology that has enabled them to be practiced in a new way, giving us a new life experience.

Digitalization Becomes the Impetus to the Global Economy

The digital industry is also integrating with offline scenes, giving rise to new business models and exploiting new tracks.

Especially in 2020 when the COVID-19 epidemic was raging, the global economy was irreversibly hit by the "gray rhinoceros," such as the U.S. stock circuit breaker, the crude oil plunge, and the sharp increase in economic downward pressure. Digitalization and intelligence have become the salvation of many industries. Strengthening technological innovation and industrial cooperation is an irreversible trend of the times.

Singapore's Minister for Ministry of Human Resources and Second Minister for Home Affairs, Yeo Lee Meng said, "Like China and other economies, Singapore is trying to turn crisis into opportunity. Let us all work together to turn the crisis into a business opportunity by leveraging on digital technology to unlock greater value, create more business models and explore new opportunities across all sectors."

At the Sands Expo and Convention Centre at Marina Bay Sands, Singapore, a mixed-reality studio has been put into use, incorporating augmented reality and hologram technology, so that even though you are in a different place, you can get the experience of being in Singapore.

Through these cutting-edge technologies, Singapore has been able to reboot the MICE industry and expand new capabilities and even reshape some important areas, making the breadth of digital-real economy integration expanding, with entity digitizing and digital entities moving in opposite directions.

In the next decade, digital intelligence technology will fundamentally change all industries that we are familiar with, such as healthcare, industrial manufacturing, education, finance, transportation and urban management. And the digitalization of traditional industries will be the new round of growth for the global economy.

In this new round of growth, digitalization is no longer a dedicated tool for large enterprises, and more and more SMEs will be empowered by digital intelligence technology to find innovative breakthroughs in digital upgrading and complete their own evolution.

The cooperation between Tencent and Zongshen Group's Kumi Network is a typical case of digital technology empowering SMEs, providing them with one-stop "new manufacturing" upgrade services around the core scenarios of Industrial Internet. For example, Taicang Technology in Chongqing has achieved transparent management of the production process through its "cloud-based factory," shortening the production cycle by 35% and reducing the abnormal loss of raw materials by 50%, greatly improving the efficiency of the enterprise.

As the precision of the integration of digitalization and real economy continues to improve, the digital upgrade of enterprises will extend from the front stage to the back stage, upgrade from the user connection to the reshaping of the production chain, from forced transformation and development to active innovation, from the aorta to the capillaries of the economy, and finally form the agglomeration effects to realize the digitalization of industry and the digitalization of the city and redefine the modes of production.

Artificial Intelligence Redefines Social Relationships

Who am I? Where do I come from? Where am I going?

Under the ultimate philosophical trifecta, what is being traced is the way people interact with the world, the relationship between people and people, the relationship between people and things, and the relationship between things and things, the sum of which constitutes our social relationships.

At present, artificial intelligence is redefining social relationships. Data resources become a means of production, computing power is productivity, and the Internet is a production relationship. These new social relationships will bring about tremendous social changes in the era of intelligence.

In recent years, artificial intelligence has developed rapidly. Take deep learning as an example, from deep neural networks to recurrent neural networks, convolutional neural networks, and then to generative adversarial networks, it is constantly upgrading and triggering new innovations.

AI continues to be the most popular technology on the list of the 2020 Gartner Hype Cycle. And several detailed categories were derived from the Hype Cycle this year, including Composite AI, Generatable AI, Responsible AI, AI Development Enhancement, Embedded AI and AI Enhanced Design.

The history of human beings is a history of interaction between human and everything, and the continuous evolution of the pursuit of low cost, low power consumption, high performance and great convenience, while AI technology holds the code of future interaction.

Artificial intelligence affects relationships and redefines them, but it won't be limited to that. The redefinition of artificial intelligence will also permeate more nodes, scenarios, time and space.

What exactly will the interaction be like in the intelligent era? Robin Li, founder, chairman and CEO of Baidu Group, said at the Smart China Expo, "In the future, every person will have an intelligent assistant."

Baidu DuerOS, a conversational AI interaction system, has been exploring the best way to interact between the human world and the intelligent world. As of June 2020, the total number of voice interactions with the Xiaodu assistant reached 5.8 billion times. More and more people are becoming familiar with and rely on Baidu's human-computer intelligence interaction system.[1]

At the same time, the interaction methods of the smart society will reduce the dependence on smartphones. Because in the era of smart industry, the types and numbers of smart terminals will far exceed those of smartphones, and the interaction methods will be decentralized from human-to-human interaction to human-to-thing and thing-to-thing interaction, from interaction with permission to interaction without permission, and from single interaction to integrated interaction.

The concept of Ubiquitous Network (UN) emerges as a reflection of the times.

Regarding ubiquitous network, experts are still unable to give the precise connotation and extension of this concept. General understanding of it is that access devices will be ubiquitous, information exchange will be ubiquitous, and converged applications will be ubiquitous.

The term "Ubiquitous Network" goes beyond the Internet and the IoT, revealing the essence of this transformation, which is no longer to influence and penetrate, but to surround and integrate. Of course, the AI brain itself does not mean anything. When the AI brain, intelligent operating system and open sharing ecology are combined together, the scenario of AI really has the power of ubiquity.

With the support of the ubiquitous network, intelligent interconnection and industry intelligence, the realistic requirement of business data is expanded to the common practice of data business. Ubiquity in connection, perception together with interaction reshapes the structure of social relationships, while artificial intelligence runs throughout the process.

Perhaps, in the next decades, people will have a better way to answer the ultimate philosophical trifecta.

---

[1] Ma Ningning, Nanfang Metropolis Daily, *Xiaodu Technology Raises Independent Financing, Valued at 20 Billion Yuan After Investment, or Will Accelerate Domestic Listing*, September 30, 2020.

Section V

Smart Dividend: China Unleashes New Growth Momentum for the World

China is known for its high quality manufacturing and the country is playing a leading role in innovation.

— Tim Cook

As the world has come to a turning point, we have opened the door to the era of intelligence. What is in the new world? Who will be the first to dig out the gold mine? We may already have some sense.

The global economy was hit hard by the COVID-19 pandemic. China quickly contained the epidemic and drove its economic recovery by upgrading smart technology, becoming the only major economy in the world to achieve positive economic growth in 2020 and unleashing new growth momentum in the world.

New Infrastructure Forges China's Economy

An epidemic disrupted the entire world, and China quickly emerged from the quagmire of the COVID-19 epidemic and moved back on track.

On March 23, 2021, the International Monetary Fund (IMF) released the latest edition of the *World Economic Outlook*, which shows that the global economy shrank by 3.3% in 2020, much worse than that during the 2008-2009 financial crisis and second only to the "Great Depression" before World War II.

Of the 17 major global economies listed in the IMF report, China was the only one to achieve positive economic growth with 2.3% GDP growth in 2020.[1]

The main driving force of China's economic growth is not only due to China's strong epidemic prevention measures and macroeconomic and financial policy support, but also due to the substantial increase in public investment in new infrastructure based on renewable energy, electric vehicles and smart industrial layout.

In 2020, China wrote "new infrastructure" into the government work report for the first time. It stated that China will "strengthen new infrastructure construction, develop

---

[1]IMF, *World Economic Outlook*, April, 2021.

a new generation of information networks, expand 5G applications, build data centers, increase facilities such as charging piles and switching stations, promote new energy vehicles, stimulate new consumer demand, and help industries upgrade." [1]

It is imperative to build a new development pattern for China to continue to achieve high-quality economic development. China noted in its consultation report to the IMF that "a substantial increase in public investment is likely to reverse the situation in the past five years, it will achieve more balanced growth." There is no doubt that 2020 was a windfall year for China in areas such as alternative fuel vehicles, the 5G chain, the semiconductor industry, pharmaceuticals and biology.

The layout of 5G is an important part of the new infrastructure. Since the issuance of 5G commercial licenses more than a year ago, the construction of China's 5G network infrastructure has been steadily advancing. By the end of 2020, basic telecommunication companies had built more than 600,000 5G base stations and 5G terminal connections had exceeded 200 million, achieving coverage in all cities above the prefecture level nationwide.[2]

China has not only cutting-edge smart technology, but also the world's largest consumer market and application market, which makes it the preferred place for AI implementation.

China has used new infrastructure-based applications and services, such as e-commerce, telecommuting, telemedicine, online education, unmanned technology, robotics, health codes and live streaming to win the "two battles" of epidemic prevention and control as well as production resumption and economic recovery. People feel more intuitively the important value and significance of promoting new infrastructure.

At the 2020 Smart China Expo Online, the Chongqing government focused on the positive effects of smart technologies applied to various industries with the support of new infrastructure.

In the Smart Construction Experience Hall at the 2020 Smart China Expo Online, digital implementation and applications at all levels from digital enterprises, digital projects to digital construction sites were displayed, including smart labor face recognition system, smart material acceptance management, Hummingbird system, BIM 5D digital project solutions, and new finance. All of them, through the combination of hardware and software, could bring immersive experience and multi-screen interaction.

At present, the development of Chongqing's digital economy has entered the "fast lane," and Chongqing is committed to build a "smart manufacturing town" and a "smart city." At the same time, because of the intelligent empowerment of big data, the city management has become smarter and smarter, and basic services have become more and more diverse.

---

[1] Xinhua News, *Government Work Report 2020*, May 29, 2020.

[2] Cui Shuang, Science and Technology Daily, *China Will Establish 600,000 5G Bases in 2021*, January 28, 2021.

By the end of July 2020, Chongqing had completed a total of 2,200 intelligent transformation projects and built 67 digital smart factories and 539 digital workshops.[1]

In the workshop of Quanta Chongqing factory, robots are the main forces on the production line. Some of them can grab parts and install them accurately in the corresponding positions, while others can hold up magnifying glasses and carefully check the errors of the products.

Similar smart manufacturing applications, smart communities, smart access control, autonomous driving, AI robots and many other scenarios are rapidly increasing both in Chongqing and in other cities in China.

World Embraces "Created in China"

The world is experiencing a new round of technological revolution and industrial change. China's accumulated advantages have burst with great vitality in manufacturing and industrial chain. In the intelligent upgrade, the IoT, intelligent cars and other areas have gained prosperous development. Intelligent products and devices have generally entered the daily life. Smart medical and Industrial Internet growth are developing rapidly, from technology followers to the role of the trendsetter in ten years.

The U.S. has been leading the frontier in basic AI research, and government agencies represented by Defense Advanced Research Projects Agency (DARPA) continue to promote AI development and applications.

Globally, China and the United States constitute the first echelon of AI, with developed countries such as Japan, the United Kingdom, Israel and France following closely behind to constitute the second echelon. Meanwhile, in the top-level design, most countries strengthen the strategic layout of artificial intelligence. At present, nearly 30 countries and regions around the world, including the U.S., China, the European Union and Japan, have released strategic plans and policies related to AI deployment.

China's AI policy is deepening into the industrialization of AI, using information technology to empower real economy and accelerating the development of "Created in China" towards high-end market. Besides, it helps the manufacturing industry achieve an industrial leap and contribute to China's manufacturing strategy.

Emerging technologies are the main engine supporting digital transformation and innovation in Chinese companies. Even against the backdrop of the COVID-19 pandemic, the landscape of Chinese technology innovation remains unique, giving new impetus to China's economic development and also giving rise to new economic industries that are driving changes in business operations and business models.

China's trend of intelligence is leading a new trend globally. Foreign governments and enterprises are strengthening their cooperation with China's new digital

---

[1]Xinhua News, *Western China Is Emerging As a New Highland of the Intelligent Industry*, September 14, 2020.

infrastructure to accelerate the development of cutting-edge application scenarios in many fields such as smart healthcare, smart education, smart culture and creativity, smart manufacturing, smart agriculture and smart transportation.

Liu Zhenmin, UN Under-Secretary-General, said at the 2020 Smart China Expo Online, "In the next decade, through effective cooperation between the government and the industry, smart technology will continue to empower the 2030 Agenda for Sustainable Development and help achieve the UN Sustainable Development Goals." [1]

In order to facilitate the flow of goods along the new international land and sea trade corridor and improve the efficiency of cargo clearance, the interconnection of the Singapore Connected Trade Platform and the Chinese Single Window has been realized, enabling the seamless flow of trade-related digital documents between the two countries, thereby reducing business costs.

In Chongqing, Intel has established the world's largest FPGA innovation center to expand the ecology for China and the world, bringing together talents and developers with a view to having 10,000 FPGA developer engineers online in 2022, providing innovation in FPGAs in the cloud and deploying the strategy in the future.

Unleashing New Growth Momentum for the World

The year 2020 has passed by, but the gloom of the epidemic has not dispersed.

The IMF has made a medium-to-long-term forecast for the global recession triggered by the COVID-19 pandemic, suggesting that the possible long-term traumatic effects will continue until 2024, while for China, the growth estimates given are 8.4% in 2021 and 5.6% in 2022. The transformation and upgrading of traditional industries through smart technology are widely evaluated as the Chinese smart dividend that unleashes new momentum into the global economy.

The Chinese government has established the goal of "basically establishing intelligent manufacturing support system, and preliminarily realizing the intelligent transformation of key industries by 2025"[2] and is working hard to provide strong support, including formulating development plans, increasing fiscal and taxation support, advancing smart manufacturing pilot demonstrations, and increasing intellectual property protection.

Opportunities such as the burgeoning intelligence process, the growing consumer market and the strong potential for business innovation provide global companies with tremendous space for growth in China.

At the "Smart New Energy—Smart New Industry" Summit of the 2020 Smart

---

[1] Yang Ye, ShangyouNews, *Wisdom at SCE|Liu Zhengmin: Intelligent Technologies Empower 2030 Agenda for Sustainable Development*, September 15, 2020.

[2] Zhao Yufei, Xuan Liqi & Wu Kunpeng, New Media (Xinhua News Agency), *China Firmly Releases "Smart Dividend" to the World*, September 18, 2020.

China Expo Online, Chen Shiqing, a member of the U.S. National Academy of Engineering and a world-renowned supercomputer expert said, "Currently, the new digital infrastructure has become an important support for the development of the digital economy." The development of digital economy in Chongqing, even China, needs to accelerate the construction of new digital infrastructure, such as 5G, artificial intelligence and intelligent supercomputing cloud platforms.

The age of intelligence belongs to the whole world, and 5G, IoT, new digital infrastructure, and artificial intelligence are not something that any single person, company, or country can accomplish on their own.

NXP Semiconductors, a leading global supplier of automotive electronics that continues to lead innovation in ADAS, in-vehicle entertainment information systems, in-vehicle networks, and automotive security, released the S32G service-based gateway processor solution at the 2020 Smart China Expo Online, marking an important turning point in the design and implementation of the entire vehicle architecture. And the reference design for the entire service-based gateway is mainly the R&D team of Chongqing Application Center.

Timothy Lee, chairman of NXP Semiconductors Greater China, said, "The development of the reference design is largely based on the needs of customers from China, and in that sense, the applications we develop locally in China are already serving the world." [1]

This is the "positive spillover effect" of China's intelligent industry, not only to enhance the level of domestic intelligence, but also to release intelligent dividends to the outside. It is vital for Asia, even for the world, and it will become a boosting engine for global economic recovery.

At present, overcoming the epidemic and recovering the economy are the top priorities of all countries in the world. China has been firmly opening up to the world, sharing the world's largest consumer market and technology application market, to welcome the coming of the intelligent era.

The gift of a rose leaves a fragrance in the hand. Everything is being connected to each other, and the concepts of a community with a shared future for mankind and the global economic development community are becoming more and more deeply rooted in people's hearts.

Both now and in the future, China will actively assume the responsibility of a great power, follow the trend of intelligent development, and contribute to the world economic recovery.

---

[1] Yang Jun, Chongqing News, *Timothy Lee: Applications NXP Semiconductors Develop Locally in China Are Already Serving the World*, October 15, 2020.

Chapter 3

Practice: The First Site of Intelligent Industry

The ultimate meaning of all changes, all innovations and all remodeling is not fanciful, theoretical and superficial, but should be practical. The intelligent industry, which has undergone many technological breakthroughs, industrial integration and commercial docking, has finally been released from the laboratory and applied in factories, farmlands, cities and various scenarios, and there are always important scenarios in which it becomes an important demonstration of this new life form of AI.

To people's surprise, AI has made the factory an unmanned smart manufacturing area that can operate automatically; AI has made the farmland a wise agricultural housekeeper that can make independent decisions; AI has made the city a wise urban life form that can evolve itself.

In this era, artificial intelligence penetrates almost all places of innovation. Even in the midst of the global COVID-19 pandemic, artificial intelligence has also acted as a hero to step forward.

Section I

Smart Manufacturing, the Transition of the Industrial Era or the Budding of the Intelligent Era ?

Science is the constant impact, breakthrough and transcendence of common sense.

— Yu Wujin

The first animal bone used for hunting, the first stone axe polished and shaped, the first dead wood drilled to make fire, the first piece of smelted and shaped bronze, the first steam engine to drive a train, the first tungsten filament to light a light bulb ...

In the long history of mankind, from the in-situ selection of primitive stone tools to the invention and creation of advanced equipment, the technological level of production tools is not only a key symbol of human evolution, but also a source of power for the progress of the times.

Entering the industrial era, the level of industrial manufacturing has become a key measure of the core competitiveness of a country, a field or an enterprise. In the development process from the industrial age to the intelligent era, the transformation and upgrading of the manufacturing industry has become an important mark of the changing course of the times.

A new chapter in the history of human manufacturing has been opened as we realize smart manufacturing step by step.

Industrial Manufacturing and Smart Manufacturing

Human civilization has a history of thousands of years, and the real start of industrial civilization is only a short history of more than two hundred years.

The loom, named Jenny, opened the curtain of the First Industrial Revolution, and in the following two hundred years, steam engines and electric engines appeared one after another, and machines became the protagonists of manufacturing.

Traditional industrial manufacturing relies more on the creative combination of tangible resource elements such as land, materials, energy, equipment, tools, capital, and manpower to enhance overall production efficiency.

In the era of smart manufacturing, all the above-mentioned tangible resource elements still exist, but the key to manifesting manufacturing capability has given way to intangible elements such as data, computing power and intelligent collaboration capability.

Through smart manufacturing technologies and smart manufacturing systems, smart manufacturing plants can perform production activities that surpass manual efficiency, even in terms of analysis, reasoning, judgment, conceptualization and decision-making, enabling more precise operation and improving overall manufacturing capabilities from a more complex dimension.

It is under this industrial mega-trend of traditional industrial manufacturing towards intelligent manufacturing that the Industrial Internet has become a new consensus in the global manufacturing industry.

At the Industrial Internet Innovation and Development Conference at the 2020 Smart China Expo Online, Li Peigen, an academician of the Chinese Academy of Engineering, elaborated on the key to the Industrial Internet, "The Industrial Internet has changed the mindset of industrial enterprises in system thinking, user thinking, collaborative thinking and ecological thinking. The future competition of enterprises triggered by new technologies will no longer happen only within the same industry, the competition in the same industry may be three-dimensional or even higher dimensional, for which the ecological thinking of enterprises should go beyond the boundaries of the industry."

In the old days, industrial machines followed the established procedures and tracks, repeating procedures day after day. In this process, small errors occurred inevitably and required manual corrections.

Today's smart machines are smaller in size and more refined in operation. There is even a "smart brain" that can automatically correct errors, prompt faults and immediately arrange maintenance robots to troubleshoot.

At the Shanghai Auto Show that opened on April 21, 2021, a high-performance electric drive coupe SUV received widespread attention due to the "Huawei official announcement" and became a new star among new energy vehicles. It is the "Huawei SERES SF5." This new car is built by the in-depth cooperation between Kingcon SERES and Huawei. It is equipped with SERES SEP200 motor + HUAWEI DriveONE three-in-one electric drive system and Huawei HiCar vehicle system. It is also the first automotive product launched by Huawei's global flagship store.

SERES Motors, which jointly built with Huawei, is actually a smart electric car brand that developed in Chongqing. It has hard cores such as "super-running power with an acceleration of 4.68 seconds per 100 kilometers" and "NEDC 1000+ kilometers of endurance." Behind the parameters, the smart factory located in Chongqing Liangjiang New District can better represent the leading level of smart manufacturing. The SERES factory takes Industry 4.0 as the manufacturing standard and is built with the goal of platform, automation, intelligence, and digitalization. It has over 1,000 robots operating in concert to achieve a high degree of automation, 100% automation of key processes and 24-hour online inspection, and the production system is fast and accurate for C2M scale customized production through big data and artificial intelligence in real-time online response.[1]

The SUPPORT P2 firefighting reconnaissance robot developed by Chongqing Qiteng Technology Company is a real "fire hero," which has powerful firefighting functions, providing a maximum flow rate of 80L/s for the powerful water cannon. It can easily drag 12 120-meter-long and 80-millimeter-wide water-filled belts to achieve rapid reinforcement of the fire scene. It is especially suitable for petrochemical, gas and other explosive environment, which is important to improve rescue safety and reduce the risks of firefighters.[2]

The applications of these smart machines and systems can produce not only industrial and consumer goods, but most importantly, provide new solutions for people.

The Qualitative Change of Smart Manufacturing Lies in the Interconnection

According to the theory of Six Degrees of Separation, each person only needs to go through six intermediaries on average to establish contact with anyone in the world, no matter which country the other person is in and what color is he or she.

Although the connection between people can be defined by a theory, it is difficult to apply the theory to the extreme in reality.

While smart manufacturing in the real world is creating a more ambitious real interconnection process. Ning Zhenbo, the former chief consultant of China Aviation Industry Group Information Technology Center shared a set of data at the 2020 Smart China Expo Online: In the development of the industrial fieldbus Internet, the interconnection of people, machines and things is most important, and the world currently has about 120 kinds of industrial fieldbuses and more than 5,000 kinds of communication protocols.

Massive industrial interconnection is realizing the connection between man and machine, machine and machine, machine and information at a speed beyond people's imagination. However, the advantages of smart manufacturing go far beyond. The immediate evolution of machine learning is updating people's understanding of smart manufacturing.

At the 2020 Smart China Expo Online, David Patterson, Turing Award winner, expressed his views on machine learning, "The first word for machine learning is 'machine,' so we need machines that learn faster and need a lot of data which are acquired thanks to the IoT. And cloud computing helps us to bring data together." [3]

---

[1]Zhang Jing, CQNEWS.com, *Shanghai Auto Show: Why Is Huawei SERES SF5 So Popular*, May 29, 2021.

[2]Liu Lei, XinHua News, *Changshou: The Mascot, YouYou, Takes You on a Cloud Tour of the 2020 Smart China Expo Online*, September 14, 2020.

[3]Yang Ye, ShangyouNews, *Turing Award Winner David Patterson: We Need Faster Machines to Promote the Development of AI*, September 15, 2020.

On the basis of machine intelligence, due to interconnection and intelligence, everything can be connected, and there is a broader application space.

Mitsubishi Electric has applied face recognition technology to elevators and developed the "ICS Intelligent Calling System." The system includes two non-contact identifications of the phone APP elevator calling and face recognition elevator calling. Passengers can use the bluetooth elevator hailing function in the mobile APP launched for elevator users to call the elevator. In the office scenario, the DOAS destination layer forecasting system equipped with face recognition avoids cabin crowding and minimizes the waiting time and elevator running time for users through pre-allocation. While in the residential elevator scenario, homeowners can also register their floors through the cabin control box equipped with face recognition.

At present, smart manufacturing has become one of the important themes of the new era, and the subsequent operation and maintenance services have a new way to start. In the experience zone of the 2020 Smart China Expo Online, China Unicom has applied 5G+AR technology to simulate a car remote operation and maintenance service scenario, where people could connect with car remote service experts on site by wearing AR glasses, and make audio and video communication based on the first-view screen to solve unexpected problems in the process of using a car in real time. At the same time, people can have a brand new experience by scanning the steering wheel though a 5G terminal to read the manual in 3D.

Smart manufacturing, in its emphasis on resource integration and value-added changes, aims to systemic changes in the industry, triggering the emergence of new models and new business forms.

Fast-Blooming Flower Buds of the Intelligent Era

Is smart manufacturing the transition of the industrial era, or the budding of the intelligent era?

This question is not difficult to answer by now. Smart manufacturing and smart interconnection have long been bridged together through AI, IoT, 5G and other technologies to breed the flower buds of the intelligent era.

Through the in-depth integration of the new generation of information and communication technology and advanced manufacturing technology, each link of the manufacturing industry including design, production, management and service has the functions of self-perception, self-learning, self-decision-making, self-execution and self-adaptation. From the initial manufacturing-oriented production to the whole life cycle of product management, till the evolution of social manufacturing, cloud manufacturing, ubiquitous manufacturing, IoT manufacturing, information physical system manufacturing and a number of smart manufacturing models with future growth potential will be formed.

Artificial intelligence began to sprout in 1956, and 50 years later, it had been endowed a brand-new strategic mission in different countries by 2008. In particular, manufacturing powers led by the U.S., China and Germany have successively proposed national strategies such as "The U.S. Advanced Manufacturing Initiative," "Made in China 2025," "German Industry 4.0" and other national strategies.

At present, smart factory and digital factory, as important practice areas of smart manufacturing, have been joined by more and more manufacturing enterprises, integrating mobile communication networks, data sensing and monitoring, information interaction and integration, advanced AI and other smart manufacturing-related technologies, products and systems. Specific applications are carried out at the factory level to realize the intelligent, networked, flexible and green production system.

Innovative pioneers such as Haier and Midea have become reference models for the construction of smart factories in various industries.

The new generation of information technology is deeply intertwined with the economy and society, and a new round of scientific and technological revolution and industrial change represented by 5G and other new generation information technologies are rapidly emerging, with unprecedented speed, breadth and depth. 5G, as a catalyst to stimulate the potential of IoT, has become the "sweet pastry" of the new generation of smart factories.

At the 2020 Smart China Expo Online, China Unicom and Changan Automobile demonstrated the whole production process of "5G+Industrial Internet" collaborative smart manufacturing factory through industrial sandbox, which utilized the integration of new generation of information technology and automobile manufacturing such as 5G, edge cloud computing and Industrial Internet to comprehensively show the integrated application of digital, unmanned and networked technologies in the production workshop, creating the first "5G+Industrial Internet" collaborative smart manufacturing factory in the automobile industry.

Driven by the new generation of information technology and advanced manufacturing technology, manufacturing companies use innovations in units, systems and management organization to optimize the production process, enhance the value of products and services and link more application scenarios.

The intelligent life we expect is that everything can be interconnected, artificial intelligence penetrates everything, and the services needed in life can be accurately observed and provided instantly by artificial intelligence. It is not difficult to see that smart manufacturing has brought the dream into reality.

On August 11, 2020, Lei Jun, in his speech on the 10th anniversary of Xiaomi, first unveiled the Xiaomi Smart Factory which is known as the "Black Light Factory." The factory uses a fully automatic production line which can produce one million smartphones a year, even without turning on the lights and human intervention. In the second phase of the planned "Black Light Factory," it may achieve 60 to 70 billion RMB output value in a year with only 100 engineers, the only staffs in the factory.

In the blink of an eye, the bud of smart manufacturing is blooming rapidly in the world with an unexpected speed.

Section II

Smart Agriculture, Redefining the Production Relationship Between People and Agriculture

Everything perfect at the moment is the result of innovation.

— John Mueller

For the whole world, agriculture is the starting point of human civilization. But when it comes to agriculture, our minds are always confined to various stereotypes such as working hard from the sunrise to sunset with only poor income.

However, relying on the development of intelligent industries, smart agriculture has begun to radiate new vitality. Through the combination of AI technology and agricultural production, people have achieved unmanned, automated, and intelligent management.

Today, smart agriculture that incorporates a new generation of information technology has become increasingly clear. Big data, agricultural robots, remote sensing technology and other technologies have long been illuminated into reality. The era of Agriculture 4.0 is imminent.

Agricultural Black Technologies at 2020 Smart China Expo Online

On February 25, 2021, President Xi Jinping, the General Secretary of the Communist Party of China Central Committee and Chairman of the Central Military Commission, announced to the world, "China has scored a 'complete victory' in its fight against poverty. The final 98.99 million impoverished rural residents living under the current poverty line have all been lifted out of poverty. All 832 impoverished counties and 128,000 impoverished villages have been removed from the poverty list. Regional overall poverty had been solved and China has accomplished the arduous task of eliminating absolute poverty, creating another human miracle that will go down in history! This is the great glory of the Chinese people, the great glory of the Chinese Communist Party and the great glory of the Chinese nation."

After the announcement of this good news, China received many congratulations from the international community. In fact, the international media has been paying attention to China's road to poverty alleviation for a long time. As early as September

2020, when China announced that it had successfully lifted more than 800 million people from poverty, foreign media reports even used the generous words of praise such as "a great feat in human history." Jeffrey Sachs, director of the Center for Sustainability Studies at Columbia University, commented, "China's struggle with poverty has been one of the most remarkable in human history and has provided experience for the rest of the world." [1]

Behind the fact that the poverty-stricken population in the rural areas of nearly 100 million were lifted out of poverty is the firm determination of a political party to eradicate poverty in an all-round way, and it is also the strong belief of a country to embrace science and technology with all its strength.

Agriculture is the gene of life rooted in the blood of the Chinese. 900 million of the 1.4 billion population have been freed from the shackles of land through decades of rapid urbanization. The transformation of agriculture carries the power of science and technology, which has explored a way for the villagers to get rid of poverty through science and technology.

At the 2020 Smart China Expo Online, many agricultural black technologies continue to emerge, relying on agricultural infrastructure, sensors and communication technology to achieve intelligence in the whole process of perception, early warning, analysis, decision-making, implementation and control, and making agricultural production more "intelligent."

In Jindai Street, Liangping District, Chongqing, there is a smart greenhouse of 35,000 square meters in the 400 acres of Digital Valley Farm. There are various intelligent agricultural cultivation modes, such as integrated irrigation technology of water and fertilizer, Dutch tomato track planting technology, bionic three-dimensional cultivation, spiral bionic cultivation, multi-layer pipelines cultivation, swing light tracing cultivation, aeroponics cultivation and abacus cultivation.

The Smartlink General Control Center is the heart of the farm. The management platform integrates modules of park management, four conditions monitoring, quality full traceability, production operation application and agricultural big data, which can realize soil monitoring, pest monitoring, safety traceability, visualization, automation and wisdom of agricultural production and operation.

Through this service platform, it is possible to observe crop growth conditions, manipulate greenhouse equipment and facilities, and guide production operations in the park in real time from anywhere with a network. It is also possible to set the automatic operation modes of the crop growth environment to ensure the best environment required for crop growth. This control center can realize the whole process of quality tracing from seeds to table, which can effectively control the quality and safety hazards in each link and provide guarantee for the high quality of products.

Compared to the traditional agriculture, information and knowledge have become

---

[1]Qu Song, Wangli, and etc., People's Daily, *A Great Feat in Human History*, September 29, 2020.

the most important elements of smart agriculture production. The two elements will be used in the production process to form knowledge productivity. The intelligent equipment will be gradually applied to the whole process of agricultural operations to achieve the replacement of human by machines, solving the problem of physical strength.

Now, with the application of the new 5G technology to agriculture, the fields have long blossomed with "wisdom flowers."

In the scientific research base of Chongqing Academy of Agricultural Sciences, a 5G networked plant protection drone can provide high-efficiency flight defense services and precision agriculture services that integrate drone plant protection, remote sensing big data and agricultural big data for experimental farmland. At the same time, the 5G network-linked plant protection UAV data will also be docked to the digital management platform of the Smart Agricultural Park of Chongqing Academy of Agricultural Sciences which will monitor field information such as soil, environment, pests and diseases in different industrial zones in the park in real time. The dynamic assessment of crop growth and efficiently-and-scientifically-guided agricultural production will be realized.

At present, Chongqing has built an agricultural and rural big data platform, including 2 systems of data standards and platform operation, 1 agricultural big data resource base, and 2 systems of government services and public services, which has initially formed a matrix of rural agricultural big data application results and promoted the innovative application of big data in the field of modern agriculture. For example, the "Yuyinong" city-wide information of every village and every household has entered into the big data platform and fully integrated public welfare, convenience, e-commerce and training services to promote the extension of information services to the village, making information accurate to the household.

At the same time, the agricultural rural big data platform also predicts the prices of agricultural products. The traditional sales channels of navel oranges were affected by the COVID-19 epidemic and the price went all the way down. Relying on the monitoring and early prediction of the platform, Fengjie navel oranges adjusted the marketing strategy accordingly to avoid the period of low prices in the national navel orange market and carried out innovative methods such as live streaming sales on the whole network and all-in-one logistics card for epidemic prevention through differentiated channels and digital marketing. In February 2020, during the most stringent period of epidemic prevention and control, Fengjie navel oranges quickly achieved a daily sale of 3,000 tons.

Smart Agriculture 4.0, Redefining Agriculture

As the first ray of sunlight reaches the ground in the morning, the irrigation system in the farmland starts to "clock-in at work." The sensors in the field also turn on the "meeting mode," exchanging the information of soil temperature, humidity and nutrients with each other, and the aggregation of data soon form the task of the day. After the meeting, the fertilizer robot automatically fertilizes the plots, while the pesticide-spreading drones start to fly over every corner of the farmland ...

Every step of a single seed, from selecting to planting, from growing to harvesting and from processing to going to market, is in the big data, visible and traceable.

Smart Agriculture 4.0 starts with the liberation of manpower. Smart agricultural machinery replaces repetitive, complicated, labor-intensive and physical labor work. Farmers don't even need to go to the field, just carry out remote monitoring.

Traditional agricultural production and operation relied on past experience, and the output of crops depended on individual efforts and natural conditions. As human society continues to progress and technological revolutions emerge, traditional agriculture has gradually evolved from the primitive model of subsistence and manual labor of the past to modern agriculture, which continuously use new intelligent systems and mechanical technologies.

As a result, individual efforts and labor are no longer major factors in determining agricultural production capacity. The strong binding relationship between man and agriculture is separated and man has become an object in the relationship of production.

Agriculture is also liberated from the limitations of individuals and no longer relies on the experience inherited from ancestors. Artificial intelligence, big data and IoT are all the results of collected human intelligence . Agronomy experts analyze the data of crops and other related factors by integrating relevant historical data such as humidity, soil quality, air indicators and weather forecasts, and find the best ratio of planting agricultural products, which thereby greatly improves the yields and quality of agricultural products.

In the future, when the operation of big data is mature, food safety regulators and agencies will also be able to collect data on food safety through online databases, the Internet, mobile smart terminals and social media. Especially in the 5G era, it is expected that smartphones will be equipped with food sensing and monitoring functions and will synchronize production records to computers and official food data centers, thus putting food safety under the supervision of all people and forming a benign regulation.

"Reinvigorating the agriculture through science and technology" has never been a simple slogan, but, in time, it should be changed. When big data and agriculture are combined, modern agriculture will be upgraded again. The Internet of Everything and big data applied to modern agriculture will show us a beautiful picture. The relationship between people and agriculture will be rewritten once again, the slogan "reinvigorating the agriculture through science and technology" will also be changed to "reinvigorating the agriculture through intelligence."

The ship of Theseus, after replacing all the old boards, has been fully transformed. Although the direction and purpose of the voyage remain the same, but it may sail faster, be more precise, and reach its destination more quickly. In a real sense, the upgraded ship of Theseus is no longer the original ship of Theseus.

Agriculture forged by technological innovation in the era of intelligent industry has also been renewed in every aspect, with the original intention unchanged and in a more robust pace.

# Section III

Smart Cities, the Fastest Transformation in the History of Urban Development

For a city, the most important thing is not architecture, but planning.

— I.M. Pei

We will eventually grow old, but the city will always be young. The city is a product of human society at a certain historical stage of development, and is one of the signs of the emergence of civilization.

Looking back at the famous cities in history, all of them have gone through a long history of civilization and industrial accumulation in units of hundreds of years. The same is true for famous cities in the world today, and most of them have their shining points in history.

The advent of the intelligent era has given a part of the cities that actively embrace the intelligent industry a brand-new opportunity for development. The rapid rise of emerging cities and the transformation of established cities have written the fastest transformation in the history of urban development.

New Smart Cities Emerge

The wave of intelligent era has swept the world, and the sparks of AI industry are scattering in every corner of the world.

Different cities follow different paths in the intelligent era. Some cities are oriented by innovative technology, some focus on industrial aggregation, and some make breakthroughs with smart applications, but they all will end up in the same way and become new stars of the city of the times.

Chongqing has taken an active part in smart city practice. Chongqing Nan'an District and Liangjiang New Area became the first batch of national smart city pilot areas in China. After several years of mapping and construction, Chongqing is now among global smart cities, achieving milestones in livelihood services, urban governance, government management, industrial integration and ecological livability.

In the construction of the smart city in Chongqing, the government has established and improved a data open service system in terms of regulations, standards, open platforms, innovative applications and cultivation of data element market by promoting the large concentration and integration of data throughout the city. In particular, since the implementation of the "Cloud Chief System" for more than a year, the "Cloud Management System" has achieved good results, and the city's cumulative data usage has increased by 168.6%.[1]

At present, Chongqing has built a smart urban management big data center, and it continues to increase the construction of the IoT application pilot programs, becoming the country's first provincial (municipal) level urban management industry big data center.

The completion rate of the urban lighting intelligent control system has reached 90%. 34 million pieces of data have been collected, and more than 20,000 video surveillance channels have been collected. Urban management has initially realized the smart services of "facility monitoring, supervision of people and vehicles, and services for benefiting the people" in one picture, one screen and one network unified management.

At the press conference of the 2020 Smart China Expo Online—Smart Applications and High-Quality Life Summit, some experts and leaders shared that Chongqing has created the first batch of 17 pilot projects of smart innovative application of "livelihood-related services" such as the one class, one-off payment, automatic reporting of road occupation, all-weather monitoring and thermal warning map. It involves the services of online education, online payment, online management, online pension and online travel, which brings new ways to enrich people's lives.

The improvement of a city's intelligence level is making the work and life of citizens more intelligent and convenient. The users of the online government management platform "Yu Express Office" have exceeded 12 million, and the number of services on the mobile terminal has reached 1,016, with more than 400 new special services for districts and counties.[2]

According to Chongqing's urban planning, the city will promote the innovative development of smart applications in accordance with the classification, so as to realize as soon as possible the sharing of smart life for all people, the coverage of the whole network of urban governance, the coordination of government services of the whole city, the beauty of the ecological livability of the whole area, the comprehensive integration of industrial improvement and the connection of the whole city infrastructure.

Chongqing is making "smart manufacturing city" and "smart city" its new "business cards," which also provides a reference for the development of global

---

[1]Xia Yuan, Chongqing Daily, *Chongqing Has Been Ranked One of China's Leading Smart Cities in 2020*, December 15, 2020.

[2]Information Office of Chongqing Municipal People's Government, *Chongqing Holds a Press Conference for the Smart Application and High-quality Life Summit of the 2020 Smart China Expo Online*, September 1, 2020.

smart cities. Artificial intelligence, big data, blockchain, Industrial Internet and other information technologies have fundamentally changed the way of business operation, people's life and public management. New expansion has been formed in the dimensions of "digital city," "information city," "wireless city," "ubiquitous city," and "smart city."

In recent years, New Songdo City in South Korea has made a global reputation. Built in 2015 on 1,500 acres of reclaimed tidal flats, it was designed as a digital city from the very beginning. The entire city's community, hospitals, companies and government agencies share information in all directions, and the digital technology reaches into every household of the city. Residents use smart cards to complete most of daily applications, including payment, checking accounts, finding cars and opening doors. Although the technology of smart cards seems to be somewhat outdated now, the penetration of digital technology is real, and the services are still shaped by replacing smart cards with mobile apps.

In addition, the Smart Civic Center in Rio de Janeiro, Brazil leads the trend of smart government. Reykjavik in Iceland, is a world leader in energy sustainability and smart solutions. Copenhagen in Denmark excels in environmental smart governance. Emerging smart cities are springing up around the world.

A New Business Card of the Old City

While emerging smart cities are standing on the tides, the globally renowned old cities are also riding the waves, implementing intelligent transformation on the basis of their original cities and changing the old into a new look with new business cards.

Essentially, a smart city is the redevelopment of a region or city using information and communication technologies to improve the performance and quality of services such as energy, transportation, utilities and connectivity to enhance the life quality of its citizens.

Singapore has long been known as a major global financial, shipping and trading center, but what is less well known is that Singapore ranks the best in the world for smart city development.

Promoting the digital upgrade of municipal services is one of Singapore's priorities in building a smart city. Users only need to make one national digital identity authentication in the system to access online services from more than 60 government agencies, including inquiries about provident fund deposits and applications for rental housing.

In Singapore, full fiber-to-the-home coverage has been achieved, and on this basis, it is committed to building a national sensor network, with thousands of sensors scattered throughout the city. Singapore wants to develop into a "smart country," and even toilets have begun to follow the high-tech route. Once the toilet emits a peculiar smell, the monitoring system will automatically remind the cleaners to clean it to ensure that the toilet is kept clean at all times.

The Silicon Valley of the United States is one of the most important technology research and development bases in the world. San Francisco is the first to benefit from this advantage. For years, San Francisco has been one of the world's leading smart cities, working to use next-generation information technology to make building operations more efficient, reduce energy use, streamline waste management systems, and improve transportation systems.

Several years ago, San Francisco replaced 18,500 light-pressure sodium streetlights with smarter and more eco-friendly LEDs. The streetlights have built-in wireless smart controllers that have the ability to monitor lights remotely, and also warn when each light burns out, thus improving safety and saving maintenance costs.

As the most well-known city label, San Francisco is the most important financial center on the west coast of the United States and the birthplace of the United Nations. Since the promotion of the smart city strategy, San Francisco has added a new business card — the greenest city in North America.

The British capital, London, has a rich historical heritage. It is the world's financial center and the city with the largest number of museums, libraries and gymnasiums in the world. Among almost all cities in the world, London is also the city with the most start-ups and programmers.

From the release of *Digital Britain* in 2009 to the *Smart London Plan* in 2013, London has been moving forward in the direction of intelligence, from smart travel to smart government services and then to smart communities, London has carried out a full range of intelligent upgrades.

At present, the London Financial City has set up all over the city with LCD digital waste recycling bins. All the bins are connected to Wi-Fi signal and can instruct residents to sort the waste disposal. At the same time, they can receive the weather, temperature, time and stock market information, effectively contributing to the construction of the smart city.

The Ultimate Vision of Smart City

What does a "good city" look like? Shakespeare answered, "The city is the man."

Regardless of the level of technological development, the development of smart cities should be human-centered. The ultimate goal is to create a better life for people in the city.

In the process of developing smart city, we should put our focus on the development of the smart society and the intelligent era. The city is just a carrier of the intelligent era, and a miniature sample of practice.

Against the backdrop of the COVID-19 pandemic, people are also full of imaginations and expectations for intelligent urban living solutions.

The cutting-edge technologies such as remote-control technology, 5G, big data and artificial intelligence had greatly helped residents and employees of enterprises who were confined to home during the epidemic to solve the inconveniences in life and work. The construction of smart cities is still ramping up.

People living in smart cities can enjoy all kinds of services at will with just a button or a signal command. The happiness, comfort and satisfaction are all close to full marks.

Every object in every corner of the city is interconnected and performs its duties in an orderly manner.

Smart street lights adjust their brightness according to the brightness of the sky. Traffic lights give the most efficient signal switches according to pedestrians and traffic flow. Smart recycling vehicles automatically take away all the city's garbage in the middle of the night. Just after dawn, watering robots send coolness to the city's green shade. Street cleaning and traffic patrol are all done by robots, and the city's weather forecast is accurate to the second…

The emergence of cities is a sign of human maturity and civilization, and it is also an advanced form of human life in groups. After thousands of years of development, human society has already upgraded the original city concept of "building a city for defense and trading," which has continuously injected new connotations into the city.

In the era of intelligent industry, the emergence of smart cities is reinventing more cities. However, no matter how the city evolves, "people" are still the masters of the city. In July 2020, the IESE released a series of conclusions and recommendations in the report *2020 City Dynamics Index* (the 7th edition). Among them, the first one is "People-oriented."

Section IV

In the Innovation Scene, Artificial Intelligence Transforms All Industries

> Every day some new miracle appears and the drama becomes real.
>
> —Miguel de Cervantes

Without a driver, cars and boats can still drive safely. When we walk into the bank, we can get a tailor-made financial plan with face recognition. When we stay in a hotel, we only need to use voice to open the curtains and electric lights, and adjust the temperature of the air conditioner. The street trash cans will be automatically reported when they are full, and they will be cleaned up in time ...

All these fantasies have become reality.

The premise for the fantasy to come true is that artificial intelligence has quietly penetrated into every corner of the world, bringing us unexpected changes.

"Black Technologies" in Life

Artificial intelligence will permeate everything!

This is not a new idea for a long time, and the trend was already indicated two years ago when Gartner Company released the 2019 Hype Cycle.

After two years of research and development, artificial intelligence has long penetrated into all walks of life, and the COVID-19 epidemic is accelerating the process. Many places such as hotels, airports, banks, scenic spots, communities, farmer's markets and public toilets suddenly become "smart."

Now when you go out to travel, you don't need to look for tips, routes and accommodations in advance. Artificial intelligence has given birth to many cultural tourism black technologies. For example, in Wulong Fairy Maiden Mountain Scenic Area, Chongqing, you can use a number of functions such as guide assistants, routes recommendation and ticket booking through a WeChat program, called "Tour Wulong on a Smartphone."

At Wulong Fairy Maiden Mountain Scenic Area, if you want to find the delicious food around you, click on the "Smart Tour Wulong" section, and then you can automatically receive the location of the nearby food. Before the trip, if you click on the "New Ways to Explore Wulong" section, the tour routes can be customized on your mobile phone. On this app, the guide assistant has voice commentary and combines real-time travel scenes, connecting the tourism consumption resources of "food, accommodation, travel, shopping and entertainment" around Wulong, and providing real-time guidance services for tourists.

All these services are linked by artificial intelligence to achieve a scenic "smart" tour.

The world's first AI multilingual virtual anchor, Xiao Qing of iFLYTEK, also "appeared" at the 2020 Smart China Expo Online and acted as the host, vividly demonstrating the strength and speed of iFLYTEK's AI voice technology.

At the expo, iFLYTEK also released its latest product, iFLYTEK Smart Learning Machine, which has brought a new revolution in education by artificial intelligence. The exhibited iFLYTEK Hearing, iFLYTEK Smart Office and other AI technology products also have been applied in many fields such as politics and law, medical care and life services, exploring the breadth of AI development from different aspects.

On the experimental roads that have been opened in several cities in China, people can already experience driverless smart taxis controlled by AI.

From 2020 to 2021, domestic innovative manufacturers of smart cars have rolled out a variety of Internet cars. The attributes of automobile products are changing from a mechanical device equipped with electronic equipment to an electronic device equipped with mechanical equipment.

In the shopping scene, AI-based customization is coming. On the AI clothing customization platform, a human body model with a specific posture can be scanned and generated through a 3D scanner. In addition, the human body reconstruction technology can automatically add bones to the scanned human body, so that it can be dressed and displayed in the virtual scene, allowing consumers to see the effects of their own dressing in the virtual space, and realizing a full range of experiential consumption and customized clothing. All the procedures of changing clothes only need one click.

The penetration of artificial intelligence, not only appears in our food, clothing, housing and transportation, but also in the most inconspicuous public toilets, presenting a new look.

In the smart public toilets built by Chongqing Zhitong Cloud Toilet Company, the temperature, humidity, odor concentration and squat usage of each toilet are clear at a glance through artificial intelligence, infrared sensing, environmental sensing and wireless transmission. In addition to bringing users an intelligent experience, smart public toilets are also practicing the concept of environmental protection. The face recognition paper feeder requires the user's face recognition to draw out paper, and only a 70-cm-long piece of paper will be provided each time. For toilet paper, if you need to take the paper again, you must take ten minutes apart, or use your mobile phone to scan the QR code. This will allow most people to save paper and also take into account the paper needs in special scenarios.

The application of artificial intelligence has opened a new era for mankind.

New Driving Forces in the Production

People's perception of life is limited, and AI technology is deeply rooted in the industrial production and urban construction, changing people's lives from the source.

Embracing change, promoting high quality development and improving business resilience have become the most pressing matters for all companies.

Among them, the smart factory construction of Chuan Kai Electric Co., Ltd. has provided a reference for the intelligent transformation of manufacturing enterprises. Schneider Electric has applied AI technology to tailor an integrated solution for Chuan Kai Electric from lean consulting to implementation, rebuilding three lean production lines, realizing a fine and flexible scheduling method, achieving reasonable distribution of materials, and reducing the quantity of work-in-process by 70%.

The transformation has reduced the seven major wastes in the production process, effectively increased productivity by 16%, significantly improved project delivery and supply chain capabilities, effectively increased warehouse utilization and reduced operating costs. With the help of Schneider Electric, Chuan Kai Electric has broken the barriers of the electrical industry, truly implemented digitalization in production and operation, achieving successful transformation and building a new path to differentiated competition.

The intelligence of smart manufacturing is not only about the factory itself, but also lies in the smart city where it is located to produce a closer linkage.

AI developers in China need a richer and more powerful AI infrastructure, including new infrastructure, new AI chips, convenient and efficient cloud services and various application development platforms, open deep learning frameworks and common AI algorithms.

Baidu's Paddle Paddle Deep Learning Platform is China's first self-developed deep learning framework, creating an "operating system for the intelligent era." Since the open source from 2016, enterprises and developers from all walks of life could use it to develop AI applications. As of May 20, 2021, the number of Paddle Paddle developers has reached 3.2 million, an increase of 70% compared to a year ago.[1]

It can be said that deep learning platforms have promoted more AI innovations, which are widely used in all walks of life, making smart cities possible.

In August 2020, Chongqing New Smart City Operation and Management Center was put into operation. It is the intelligent hub of the smart city. Its data resource center gathers all kinds of governmental data resources in Chongqing, and now accesses a total of 43 systems from 21 departments, districts and units, including Chongqing Municipal Government Office, Chongqing Municipal Ecological Environment Bureau, Chongqing Municipal Health Commission and Chongqing Municipal Urban Management Bureau, involving online management, online services, online business and other important

---

[1]Lin Zhijia, TMT Post, *The Number of Paddle Paddle Developers Has Reached 3.2 Million, 70 Percent from a Year Earlier*, May 20, 2021.

livelihood-related applications.[1]

The smart hub explores deep into the value of data, monitors early warning center to realize the perception of the city's operational situation, monitoring analysis and predicting early warning, and schedules the command center to realize the supervision on daily work and urban emergency handling.

The smart hub can comprehensively enable the platform to provide generic technology, business cooperation, security operation and other support services, constituting the city's "brain." One network of unified management, administration, scheduling and governance will become a reality.

Artificial Intelligence in the Post-Epidemic Era

In the post-epidemic era, the world is facing great uncertainty, both economically and socially.

Fortunately, with the help of a new generation of information technology and the blessing of global digitalization and intelligence, the whole world is also taking advantage of the opportunities of the intelligent industry era to continuously reduce these impacts.

In the past year when the whole world has been fighting against the COVID-19 epidemic, AI technologies such as face recognition have played an important role in public health, such as modeling and prediction of the whole epidemic, tracking, detection and data analysis of the epidemic, and other services like unmanned food delivery and medical devices.

"AI + medical" is one of the future trends of artificial intelligence. Comprehensive health, medical diagnosis and new drug development have benefited from the development of AI technology. For example, in terms of the study of genetic mechanism and the new tools of high-throughput histology, AI may play a vital role.

In addition to the medical field, what other aspects of AI will make a breakthrough in the post-epidemic era? Those industry elites attended the 2020 Smart China Expo Online brought a whole new set of thoughts.

Robin Li, founder, chairman and CEO of Baidu Group, thought that digital intelligent technology would fundamentally change all walks of life as we know it, such as health care, industrial manufacturing, education, finance, transportation and urban management. Currently, Baidu has deployed in the fields of deep learning, autonomous driving, intelligent interaction and AI platforms.

AI technology will serve for industrial intelligence. The future of industrial intelligence will eliminate traffic congestions, improve production and work efficiency, reduce resource waste, realize people-friendly smart services, and establish a more civilized and secure smart society, which is exciting and expected.

---

[1]Huang Xing, Xinhua News, *The Intelligent Hub of the Smart City Was Put into Operation in Chongqing*, August 22, 2021.

Peng Honghua, president of Huawei's 5G product line, believed that the hottest 5G technology right now would catalyze everything. 5G would connect the whole scene and realize the Internet of Everything. 5G would make artificial intelligence omnipresent, make computing ubiquitous and cloud accessible, accelerate the digitalization of millions of industries, and create new values and opportunities.

Meng Pu, chairman of Qualcomm China, believed that the combination of "5G + AI + edge cloud" would transform education, accelerate medical and automotive industry transformation, improve work efficiency and build flexible manufacturing systems. At the same time, it could also build a borderless extended reality ecology and realize personalized shopping experience.[1]

Most importantly, the X effects brought about by intelligence would bring great changes to many industries.

Xu (Ian) Yang, corporate vice president and president of Intel China, believed that the "smart X effects" would bring more digital value-added services[2], especially after the epidemic, online applications such as online office, online learning and online shopping would be transformed from temporary applications to rigid needs of life.

In the face of such a huge structural transformation of the economy, the support of key underlying technology would be particularly important, and the uncertainty of the future would be traceable.

[1] Xie Yiguan, China News, *China Has Built More Than 500,000 5G Bases to Promote the Development of Converged 5G Application*, September 16, 2020.

[2] Tencent, *Xu (Ian) Yang: Smart X Effects Would Bring More Digital Value-added Services*, June 23, 2020.

… Chapter 3 Practice: The First Site of Intelligent Industry

Section V

China Adopts AI Practices to Fight Against the Epidemic Under the Pandemic Dilemma

Practice will solve those problems that cannot be solved by theory.

— Ludwig Andreas Feuerbach

The core driving force for the sustainable development of human society comes from the ability of human beings to solve problems. When encountering any problems, it is one of human instincts to actively seek solutions.

When human beings have artificial intelligence, those once complicated and helpless problems are being solved at a faster rate.

In 2020, the COVID-19 epidemic outbreak became a global black swan incident, and China took the lead in practicing anti-epidemic by AI technology, which has provided many valuable lessons to the world.

The Rise of Contactless Services

There has never been a time when people are more overwhelmed than in 2020.

People have never thought that the way to contact an elevator button could be so diversified and creative. Daily necessities such as tissues, toothpicks, cling films and sticky notes became the magic weapons for people to avoid direct contact with the elevator button, but still did not solve the problem radically.

For example, Baidu, together with Hisense Real Estate, Country Garden and other partners, launched a contactless elevator ride solution with voice calling, using AI voice recognition to achieve zero contact elevator control, making mobility more convenient and safer. This solution is currently applicable to most elevators in the market, while compatible with the original elevator control system.

In the early days of the COVID-19 epidemic, China adopted a strict "lockdown" policy to strongly curb the spread of the epidemic. Most of the country's 1.4 billion people were confined to their homes, but life went on. The COVID-19 epidemic has given rise to a large number of contactless services, including home office, online education and online grocery shopping.

This unexpected epidemic has brought huge and severe challenges to the anti-epidemic work in catering and leisure industries, supermarkets and convenience stores, culture and entertainment facilities, and tourism across the country. It was not enough to rely on manpower to win the battle against the epidemic.

Especially in ensuring people's daily life, courier companies and takeaway companies acted quickly and timely to promote "contactless delivery." Delivery robots, unmanned logistics warehouse, smart cloud warehouse system and other technological solutions were put into use.

In Wuhan, when the epidemic was most serious, Jingdong Logistics quickly completed the map data collection and testing for robot delivery, arranging delivery robots from all over the country to support Wuhan. Logistics drones, which have been in the R&D testing stage, also tried to come on board in advance to provide logistics services for the lockdown areas.

In Chongqing, the Municipal Commission of Commerce, in conjunction with Meituan Dianping, launched a directory of 249 food and beverage stores to fully implement the "contactless ordering" service, and increased the supply of food and beverage takeaways to provide safe food for the residents and employees of enterprises returning to work and production.[1]

For small-and-medium-sized enterprises to resume work and production, technology companies used cloud computing to vigorously promote enterprises to the cloud, focusing on online work methods such as telecommuting, home office, video conferencing, online training, collaborative R&D and e-commerce. Using "Internet + technology" to maintain a contactless office was an important strategy to fight against the epidemic and sustain the economy, and would also bring about an overall change in the way offices operated in the future.

The online recruitment has become the preferred form for enterprises to attract talents. All recruitment activities were transferred to online practices, so that employment services, online recruitment and interviews were conducted through online voice and video.

In addition to satisfying people's basic needs for food, clothing, housing and transportation, contactless services have influenced various industries. The financial sector, which is the binder for the resumption of work and production, is also being reshaped in the midst of turmoil. As a provider of financial services, the banking industry, on the basis of fully integrating financial technology, has applied artificial intelligence, big data, cloud computing and other technologies in contactless services. It has greatly reduced or avoided direct contact with customers to develop a new generation of financial services, to accelerate digital transformation, to integrate new business scenarios and to enhance the service capabilities of "contactless finance."

AI technology has strongly promoted the cooperation between manufacturing

---

[1] Huang Guanghong, Chongqing Daily, *249 Food and Beverage Outlets Implement the "Contactless Ordering" Service*, February 18, 2020.

enterprises and information technology enterprises, deepened the application of new technologies such as Industrial Internet, industrial software, artificial intelligence and augmented reality/virtual reality. AI technology has also played a vital role in promoting new models and business forms such as collaborative research and development, unmanned production, remote operation and online services, and accelerating the restoration of manufacturing capacity.

The Highlights of Technology Against the COVID-19 Epidemic

On February 4, 2020, the day after the strictest prohibition order was issued, the Yuhang District Committee Office of Hangzhou held a meeting and clearly proposed a set of digital solutions to achieve three responding plans of "full population coverage + full process control + full field joint defense."

At 5:00 a.m. on February 5, 2020, the first version of the solution was born, after which the optimization was made every half hour. Through cell phone positioning, operator data and big data comparison of epidemic dispersion, it could present the epidemic prevention health information of the code holder with QR codes in red, yellow and green colors, and provide other subsidiary functions.

The transformation process of Yuhang District from the "hardest-hit" area of epidemic in Hangzhou to being the first to resume production took only one week. This was credit to the set of digital solutions developed based on AI technology.

Subsequently, Yuhang's trial experience spread nationwide within 40 days, which is what we all know as "Health Code." Whether it is to enter and exit various cities, to enter and exit the community, or to pay for medical treatment, the health code is the most effective pass credential.

In various cities, health codes have been evolved in line with the local epidemic prevention. In Chongqing, Tencent has combined the "Yu Kang Code" and the "Ride Code" into one, evolving them into the "Health Ride Code," which realizes the "One Code Passage" between various modes of transportation. The one-off code can be used to complete health verification and fare payment in one swipe, making mobility more convenient.

During the epidemic, the "Health Code" has become a necessary "Exit Ticket" and "Pass" for 30 million Chongqing citizens.

The key reason why China was able to resume work and production as quickly as possible in the world was the innovation of temperature measurement technology in public places, especially in densely populated areas such as hospitals, railway stations and airports. The test results must be more accurate, and the test efficiency must be higher as well. In order to cope with the fever of hospital patients and peak return passenger flow, 5G technology has also come in handy. Many areas have used 5G thermal imaging temperature measurement systems in hospitals and train stations.

In this case, 5G technology has also been put to use due to its fast speed and low latency. Many regions like hospitals and train stations have adopted 5G thermal imaging temperature measurement systems. Through the combination of infrared body temperature detection camera and 5G wireless access equipment, the body temperature can be detected within 10 meters in seconds without contact.

Baidu AI has also developed face key point detection and image infrared temperature dot matrix temperature analysis algorithm to detect the forehead temperature of passengers within a certain area, even if they are wearing hats and masks. In scenes such as subways and high-speed trains that require a large amount of body temperature monitoring, more than 200 people can pass through a single channel at the same time within 1 minute. During the detection process, passengers hardly need to stop, and the temperature recognition error is only ±0.3°C. Such non-contact temperature measurement has greatly reduced the risk of cross-contamination.

In the 5G era, medical forces around the country could save lives and help the patients when they were not at the front line. During the epidemic, the First Affiliated Hospital of Kunming Medical University enabled the first 3D digital teleconsultation system with AR/5G. Even if the patient was thousands of miles away, as long as the doctor wore VR glasses, a 3D COVID-19 patient's lungs would display in front of his eyes, and the lung lesions would be clear at the first sight.

Previously, Huawei donated a telemedicine platform for Huoshenshan hospital, which directly connected to medical experts in Beijing and Shanghai. The solution consists of Huawei TE20 video conferencing terminal + 5GCPE + Smart Screen + Huawei Cloud, and the system supports 1080P HD picture quality.

Huawei Cloud WeLink is now open for "telemedicine solutions," including the provision of remote diagnosis and treatment, remote visitation, remote conferencing, case collection and targeted notification. Through "remote diagnosis and treatment" and "remote visitation," experts and doctors can conduct a full range of high-definition remote consultation, timely diagnosis and treatment, and the public can also remotely visit their families through video conferencing to avoid the risk of infection.

There are no national boundaries in the fight against the COVID-19 epidemic. Facing the global public health crisis, Chinese companies are taking actions, exporting detection and diagnosis capabilities overseas through artificial intelligence, and contributing to global cooperation against the epidemic and economic recovery.

In the Post-Epidemic Era, Artificial Intelligence Is Still the First Mainstay

People are always full of expectations for artificial intelligence, and how to implement it has always been the key to the development of AI innovation projects.

In this COVID-19 epidemic, AI applications basically covered all aspects of the whole process, especially in epidemic monitoring and analysis, personnel and material control, logistics, drug development, medical treatment, and resumption of work and production. With a large number of landed applications, labor costs are greatly saved, human resource consumption is reduced, efficiency is improved, and the risk of virus infection transmission is greatly reduced.

In the early stage of technological battle against the epidemic, although AI was put into use in large quantities, the actual application was superficial and did not meet people's expectation of AI. In community prevention and control as well as front-line work, human resources were still the absolute main forces.

The sudden outbreak of the COVID-19 epidemic has caused insufficient data accumulation. AI technology has made mediocre performance in the detection of virus transmission, spreading routes and the traceability of the source of the virus, and has not played its due role.

Although some big data and AI tools have been used in the process of vaccine development, the front-line medical personnel and scientists doing pharmaceutical research and vaccine development have made the main contributions.

In people's expectations of artificial intelligence, "AI + medical" is a general direction, but it is clear that there is still a lot of space for the development of AI in the medical industry, such as the prediction of protein structure, the study of the gene mechanism and the exploration of high-throughput sequencing[1].

In the future, AI will be adopted at an accelerated rate in many other areas of healthcare, and AI technology will be used to deal with a large backlog of other medical problems such as cancer, not just to deal with the spread of viruses.

The COVID-19 epidemic has had a huge impact on the way we lived, worked and socialized. A steady and strong digital trend, which was already visible previously in many aspects of society, was setting off a digital boom worldwide in 2020.

In the post-epidemic era, technology is evolving at an astonishing rate, constantly defining and updating the next frontier, and AI remains the first mainstay of the technologies that are changing the way we live, work and entertain.

In the future outlook of the Internet of Everything, smarter big data analysis, more timely detection and prevention, customer stickness, and smarter decision-making capabilities are the most urgent goals of AI. Cognitive intelligence technology and emotional computing will both usher in new demands, and human-machine coexistence is close at hand.

The developing artificial intelligence is still far from global popularization, but it does not stop us from imagining and looking forward to it.

---

[1] High-throughput sequencing, also called next generation sequencing (NGS), is massively parallel, sequencing millions of DNA fragments simultaneously per run and employing "short-read" sequencing.

Chapter 4
Achievements: Cutting-Edge Innovations in Technological Intelligence

Different scenes need different innovations; different changes create different achievements. The exploration of various fields, the solutions to many problems, the repeated experiments and crossover connections have made artificial intelligence walk from the streets to the square, and finally gather on the stage of the intelligent era.

Smart materials, industrial interconnection, autonomous driving, smart cultural tourism, smart epidemic prevention, smart finance, smart government... When we review the AI innovations from different fields one by one on the stage of the Smart China Expo, we realize that many AI innovations together hold up a brand-new era.

Section I

The Convergence of Black Technologies, Overlook the Future of the Times from the Smart China Expo

Catch up with the future, capture the essence of it, and change the future for now.

—Nikolay Gavrilovich Chernyshevsky

Let the future come now is the greatest expectation of the whole world. With all the different technology product launches being held around the world every day, these new technologies feel like a collective creation of the future by technology innovators around the world. However, not everyone is able to piece together the whole picture of the future with these scattered fragments of innovation.

The Smart China Expo, which began in 2018 and has been held for three consecutive sessions, brings together global technological innovations, like a future-themed feast that attracts people to "taste."

At every Smart China Expo, new technologies and new ideas are emerging. The cutting-edge technologies have promoted all areas of society to accelerate the progress of intelligence, and also created a new window of synchronous communication between humans and the future world.

The 2020 Smart China Expo Online gathered talents, attracting a total of 551 domestic and foreign enterprises in over 20 fields including 5G, blockchain and Industrial Internet, and held a total of 41 forums. 8.7% of the world's top 500 companies and 10.3% of China's top 500 companies were involved at the online expo.

Smart factories, 5G remote driving, smart materials and dual-wrist robots showcased at the 2020 Smart China Expo Online. A number of new technologies and products such as silicon photonics technology, L4-level self-driving mini bus and 8K Micro-LED also made the first debuts at the expo. The ten Industrial Internet identification resolution secondary nodes covering medical devices, automobiles and other fields were officially launched.

Compared with the previous two sessions, the achievements of the 2020 Smart China Expo Online have been extended on a large scale in terms of applications. Many of the innovations have been applied to the market from the laboratory. Technologies have already penetrated into people's daily lives and are no longer out of reach.

These cutting-edge technologies from around the world have become the barometers of future technology development.

The new generation of AI breakthroughs and applications has further enhanced the level of digitalization, intelligence and networking in manufacturing, and accelerated the speed of innovation and promoted a wider range of applications.

At the 2020 Smart China Expo Online, three Nobel Prize and Turing Award winners, 49 academics, 443 renowned experts and industry elites collided ideas and exchanged achievements around frontier hot topics such as artificial intelligence, 5G, blockchain, and Industrial Internet. New ideas, new concepts and new perspectives were dispersed globally, stirring up unrestrained ideas about the future.

In the near future, flying cars, brain-machine interfaces, the empty but busy factories — scenarios that seem to be sci-fi today — may become commonplaces. In the next 50 years, human life will be reshaped by artificial intelligence, and all the changes will find their sources in today's innovations.

Machines are becoming smarter, systems are more fluid and accurate, ubiquitous connections are everywhere, and the process of the intelligent era has become irreversible. Machines are becoming more and more similar to people, and even perform much better than people. The only thing people have to do is to continue to use their imaginations and focus their energy on more meaningful explorations of the unknown.

Perhaps, in the future, everyone will be a programmer or an innovator. Technologies and machines are allowed to be transformed by our creative thoughts.

Section II

From Smart Materials to Autonomous Driving, Manufacturing Is Changing at an Accelerated Pace

Don't be afraid of new arenas.

— Elon Musk

❝High intelligence" products can not be separated from the "high intelligence" manufacturing.

In the new round of technological revolution and industrial transformation, smart manufacturing has become the commanding height and main direction of development opportunities for countries all over the world.

With sensors, IoT, big data, cloud computing and other new technologies, the industrial manufacturing industry is ushering in a moment of change. When the factories are connected to the cloud, products have brains.

Research and Development of Smart Materials

In the past, mechanical materials and structures were lifeless. Now, mechanical structures have a nervous system by affixing "sensors," and mechanical structures have muscles by affixing "actuators." Lifeless machinery seems to become "flesh and blood" life. All these changes are attributed to the role of smart materials.

Smart materials, also called mechanically sensitive materials, are new functional materials that can sense external stimuli, judge and process appropriately, and can perform themselves. They combine high-tech sensors or sensitive elements with traditional structural and functional materials to give materials new properties, making inanimate materials seem to have "feelings" and "perceptions" and have self-awareness and self-healing functions.

On the opening day of the 2019 Smart China Expo, Chongqing Nobel Prize 2D Materials Institute was officially registered and established. It is committed to basic research, industrial projects and standardization and other industrial services. Professor Kostya Novoselov, the winner of the Nobel Prize in Physics in 2010 and one of the discoverers of graphene, serves as the honorary dean of the Institute and leads the team.

By the time the 2020 Smart China Expo online was held, the Institute had assembled a research team of more than 20 people in less than a year. At the 2020 Smart China Expo Online, Kostya Novoselov presented the Institute's latest research results in smart materials.

The smart materials are the fourth generation of materials after the natural materials, synthetic polymer materials and artificially designed materials. They are one of the important directions in the development of modern high-tech new materials which will support the development of future high technology and erase the boundary between functional materials and structural materials in the traditional sense gradually, realizing structural functionalization and functional diversification. Scientists predict that the development and large-scale application of smart materials will lead to a major revolution in the development of material science.

Within just two years, the basic research on 2D materials has already achieved major breakthroughs. With the help of innovative growth technologies, the Institute has been conducting research on the production of new 2D materials, not only to achieve pure 2D crystal growth, but also to take into account alloy growth, and try to combine 2D materials with other materials and crystals, such as polymers and polymeric electrolytes. Therefore, these materials will have a wide range of properties and can cope with more different challenges and tasks in the intelligent era.

In response to COVID-19 epidemic, the Institute has embarked on a study of intelligent anti-viral coating that will be able to kill viruses adsorbed on surfaces through an intelligent programmable response, which is highly effective for many viruses, including the new coronavirus.

However, that's not the end of it. The Institute will extensively use machine learning technology in the field of material science and study 2D material crystals and their synthetics. Besides, the Institute will further improve these new materials and produce some sort of programmable materials.

Currently, the Institute has invested a lot in smart textiles, thermal management, telecommunication, photonics, surface plasma photonics and other applications, and has performed well in smart membranes and smart coating.

Adaptive smart materials that meet a wide range of functional needs and adapt positively to the external environment will find applications in many fields such as smart cities, robotics, artificial intelligence, advanced healthcare, water treatment and telecommunications.

Kostya Novoselov said, "We want our materials to be alive, to function like living systems. We believe that our adaptive smart materials will find applications in many fields such as smart cities, robotics, artificial intelligence, advanced healthcare, water treatment, and telecommunications." [1]

---

[1] Yang Ye, Shangyou News, *Kostya Novoselov: The First Trial of Intelligent Antiviral Coating Against COVID-19 Has Achieved Promising Results*, September 15, 2020.

Development of Industrial Internet

For some time, the era of Industry 4.0 has been guiding the direction of manufacturing transformation. In the meantime, the Industrial Internet is highly expected. The Industrial Internet identification resolution system has played a major role in the development of the Industrial Internet.

At the 2020 Smart China Expo Online, the China Academy of Information and Communication Research (CAICR) made a comprehensive exhibition of the national top-level node for identification resolution system. Back on December 1, 2018, China's national top-level node for Industrial Internet Identification Resolution was launched in Chongqing. This national Industrial Internet infrastructure is expected to accelerate the pace of development of Industrial Internet in the western region and enhance the capability of basic services.[1]

In simple terms, Industrial Internet identification decoding is the "identity card" that issues unique coding for many networked machines, equipments and sensors. Just like each of us has a resident ID card, with the identification resolution system, it is possible to uniquely locate and query information of machines, equipments and sensors to realize the accurate docking of global supply chain system and enterprise production system, the whole life cycle management of products and the implementation of smart services.

The national top-level node of identification resolution is the top level within a national or region. It can provide basic services and resource management functions of domain names, logos and blockchain for the whole country. The national top-level node should be connected to both the international root nodes of various identification systems and various other identification nodes at the secondary and below in China.

As the top identification service node in China, it can provide top-level identification resolution services and management capabilities such as identification filing and identification certification for the whole country, and connect the international root node and the secondary node to achieve connectivity. It can be said that the national top-level node occupies the position of infrastructure in the whole Industrial Internet standard resolution system.

As of the end of May 2021, China's Industrial Internet identification registrations have ushered in "bursting nodes." A total of five national top nodes have been established and respectively located in Beijing, Shanghai, Guangzhou, Chongqing and Wuhan. 134 secondary nodes have been launched, covering 28 industries in 23 provinces, autonomous regions and municipalities. With more than 15,000 access enterprises, the total number of national logo registrations has exceeded 20 billion, and

---

[1] Yu Hongtong & He Zonghan, Xinhua News, *China's National Top-level Node for Industrial Internet Identification Resolution Was Launched in Chongqing*, December 3, 2018.

the average daily resolution has reached 12 million times.[1]

Chongqing's top-level node went online in December 2018 and became an important new infrastructure for the digital economy. According to the Institute of Industrial Internet and IoT of China Communications Institute, as of May 21, 2021, the total number of national top-level node (Chongqing) identification registrations reached 631 million, the cumulative total number of identification resolution reached 334 million times, and 19 secondary nodes and 1,099 enterprise nodes have been accessed.[2]

The Industrial Internet has brought China's manufacturing industry into a new era, formally introducing "personalized customization" to manufacturers and consumers throughout China, and the C2M model has become the new market opportunity.

In the C2M production and sales collaboration model built by Chongqing Feixiang Industrial Internet Co., Ltd. and Ali Cloud, mass production and personalized customization are dovetailed. The platform links business partners including end customers, distributors and flagship stores through Taobao sales system to obtain orders directly and realize the information collaboration between Taobao factories and the sales terminals.

In the C2M model, from the initiation of personalized demand of the consumers to the delivery of the final products, there are many stages in between, including product design, R&D, manufacturing, logistics and services, and involving multiple scenarios from the consumer platforms to the manufacturing plants. without the Industrial Internet in which connectivity and synergy play a full role, C2M would be nothing but an empty talk.

The massive user orders are transmitted back to the factory in real time through the Industrial Internet platform, and the information is automatically transmitted to each workstation, process, production line, and each supply chain link to achieve synchronized production and finalize the delivery of personalized products based on MES intelligent decision-making and automatic production arrangement.

Tao platform quickly processes orders, intelligently generates planning orders, synchronizes the procurement of the parts supply chain, accurately issues production orders according to inventory, equipment status and order scheduling, and monitors the production process in real time. Smart logistics achieve smart warehousing, real-time monitoring of transport, customer sign-off, reconciliation and settlement, forming a closed loop of data.

Moreover, users who place orders can monitor the production and progress of products in real time through cell phones, computers and other terminals to realize the visualization of production and manufacturing.

---

[1] Huang Sheyu, Paper CNii, *The Total Number of National Logo Registrations Has Exceeded 20 Billion*, June 1, 2021.

[2] Xia Yuan, Chongqing Daily, *The Total Number of National Top-level Node (Chongqing) Identification Registrations Reached 631 million*, May 30, 2021.

The Achievements of Automated Driving

Since Google announced in 2010 that it was developing self-driving cars, the number of followers has been growing day by day, with Apple, Baidu, Ali and many other tech giants vying to be the first to start research and development.

Today, 11 years later, new and traditional auto enterprises have also joined these technology companies. Nio Auto, Li Auto and Xpeng Motors in the field of electric vehicles (EVs) have launched their electric cars one after another, and besides, Xiaomi and Baidu have officially announced the production of electric cars.

In China, Baidu announced the development of self-driving as early as 2014. After 6 years of trials and accumulation, Baidu unveiled the Robobus, an L4-class self-driving bus, to the public at the 2020 Smart China Expo Online.

Once the self-driving bus was unveiled, it astonished the audience. The bus is 5.9 meters long, with an approved load of 19 people. Red-painted body, "big eyes" on the head, and "wind ears" on both sides all come with an aura of technology. The "big eyes" are the monocular cameras of the self-driving bus, and the "ears" on both sides are LIDAR. The vehicle is equipped with a total of 4 lidars, 2-millimeter-wave radars and 7-monocular cameras. These devices help the vehicle detect all kinds of information in the process of driving, and the omnibearing monitoring makes its sensing more comprehensive.

When the vehicle is driving normally, it can automatically identify and stop to yield to pedestrians or vehicles when they encounter a crossing. In addition, the self-driving bus can easily identify traffic lights and realize the operations like moving straight ahead, turning and turning around. Furthermore, it can notice all the situations in front and behind during the automatic driving process.

In order to adapt to the distinctiveness of the bus scenarios, Baidu's Robobus has additional designs in terms of stops and throughput combined with realistic usage, with precise parking capability to achieve accurate stops, easily coping with bus stop scenarios and more complex urban road conditions.

The self-driving bus also has a deep learning function. When it repeatedly drives on the same road every day, it is able to analyze and record the road conditions and analyze accident-prone sites, thus enabling the car to anticipate possible situations in advance.

According to the plan, the first self-driving bus line in western China, located in Yongchuan District, was operated in September, 2020. Chongqing people have experienced riding on Baidu's Robobus, a self-driving bus, and Robotaxi—Apollo Go, a self-drving taxi.

Besides, Chongqing Yongchuan District People's Government, Chongqing Vehicle Inspection Institute and Baidu jointly built the "Western Automated Driving Open Test Base," which has been officially put into use.

The Western Automated Driving Open Test Base is the open test and demonstration operation base with the richest application scenarios and the largest scale of automated driving vehicles in western China, and the first L4-level automated driving open test and demonstration operation base in China. The base is located in the central area of Yongchuan, Chongqing. Based on the construction mode of comprehensive coordination of vehicles, roads, clouds and maps, the base has deployed 5G communication road network environment in all aspects and built more than 30 typical open road test scenarios in mountain cities such as interchanges, tunnels and bridges, which can carry 200 intelligent driving vehicles for testing at the same time.

Currently, Baidu has been granted 10 licenses to operate self-driving manned tests in Chongqing, and has conducted tests and demonstrations of 20 L4-level self-driving vehicles in five application scenarios at the base, including the self-drving taxi jointly developed with Hong Qi, Ford and Lincoln, the self-driving bus developed in cooperation with Xiamen King Long and the driverless sanitation vehicle built in cooperation with Qingling Automobile.

The base also has fully opened the Baidu Apollo autonomous driving test cloud control platform, which has realized the functions of road network environment perception, roadside edge computing and real-time vehicle-road information interaction. This platform has fully integrated the national motor vehicle testing capability platform of Chongqing Vehicle Inspection Institute, and taken the lead in western China to build a "virtual simulation + closed experiment + open test" as a whole. The company has been approved as one of the top ten application scenarios in Chongqing National New Generation Artificial Intelligence Innovation and Development Pilot Zone.

The Western Automated Driving Open Test Base, as a leading intelligent networked innovation sample in the west, has accelerated the take-off of China's artificial intelligence and intelligent transportation, and provided good R&D testing services and rich demonstration application scenarios for domestic and foreign autonomous driving automobile enterprises, system solution enterprises and auto parts enterprises, jointly calling for the coming of the intelligent era.

Smart Manufacturing Makes Products Smarter

In the new intelligent era, people have enjoyed unprecedented freedom, both spiritually and materially.

The footnote of freedom has been written by cutting-edge technologies like artificial intelligence, 5G and IoT. The new generation of information technology has brought people faster logistics, more efficient services and smarter products, and the old products and services have been boosted by technology to attract people's amazed eyes with new faces.

In the old Chongqing people's eyes, Duan's Suit was full of the flavor of the old time, and the new world seems to have no place for it. But after meeting the artificial intelligence, Duan's Suit has regained its glory.

Chongqing Duan's Clothing AI customization platform has made fitting more intelligent. The 3D scanner can scan and generate a specific posture of the human body model and the human body reconstruction technology can automatically add bones for the scanned human body so that they can virtually dress and display. Moreover, the fabric simulation machine is added to quickly mesh the 2D clothing pieces and display the same physical characteristics as real clothing pieces in a 3D environment. A computer graphics base can show the same effects of folds and bumps as real clothes worn on real people.

The fitting in this customization platform has no difference from the fitting in the physical store. The artificial intelligence could help Duan's Suit identify the defects of the dressing effects, rapidly modify the version, quickly cut out the most suitable clothing for the consumer's body size so as to achieve a full range of experiential consumption and personalized clothing customization.

From smart materials to Industrial Internet and then to industrial robots, the manufacturing industry has ushered in a chance to take off, and a new generation of display technology — 8K Micro LED has also come into being with the help of smart manufacturing.

At the 2020 Smart China Expo Online, Chongqing Konka Optoelectronics Technology Research Institute Co., Ltd. showcased its 236-inch 8K Micro LED TV which is the above-100-inch 8K MicroLed display products with the world's highest image resolution. There are more than 99 million areas for light control and peak brightness of 3000 nit, contrast ratio of 1000000:1, and equipped with the world's most advanced driving solution + 5G communication module.[1]

The decision of Fengmi Technology to settle in Chongqing is also based on Chongqing's vigorous cultivation of the "chips, screens, smart terminals, core elements and IoT" industry chain, which is consistent with Fengmi Technology's main business of ALPD laser display technology.

At the 2020 Smart China Expo Online, Fengmi Technology showcased the "Fengmi 4K Laser Cinema Max." The product has a picture brightness of up to 4500 ANSI lumens. The leading HDR10+ and HLG decoding technology bring a more delicate picture contrast. In addition, the 0.25:1 large depth of field and ultra-short focal lens design can easily project a giant screen of up to 200 inches, restoring more realistic cinematic experience, and truly let viewers "move the theater home."

The new generation of information technology is performing well in the field of energy beyond the applications in our daily lives. China Shipbuilding Industry Corporation Haizhuang Wind Power Co., Ltd. independently developed the LiGa smart operation and maintenance platform for onshore and offshore wind power equipment, which can realize smart wind power remote operation and maintenance and make

---

[1]Liu Zhengning & Zeng Qinglong, People's Daily, *Bishan Enterprises Will Appear at the 2020 Smart China Expo Online with Black Technologies*, September 12, 2020.

energy smarter.

On the LiGa platform, a large amount of wind farm operation and maintenance data, meteorological data and wind farm management experience produce chemical reactions.

LiGa platform can make intelligent decisions on overall solutions such as wind resource assessment and macro site selection, big data collection and storage management, wind farm intelligent operation and maintenance, wind farm post-evaluation, data mining and analysis, fault diagnosis and predictive modeling analysis. It can also realize remote real-time monitoring of wind farm operation status, smart health management of the life cycle of the wind turbine, product analysis and assisted decision-making.

As 5G technology gradually matures, remote services have a solid support. Several manufacturing companies are exploring 5G remote operation and maintenance. 5G smart factory, 5G remote driving and other businesses, smart manufacturing, smart agriculture, smart medical care and smart transportation are emerging.

It has become a common strategic choice for all countries in the world to develop intelligent industries with AI technology as the core driving force and cross-industry integration as a typical feature, and to promote economic and social transformation and upgrading.

In the field of industrial manufacturing, this choice has long been a global consensus.

Section III

New World at the Smart China Expo Empowers Life to Leap "Intelligently"

Science does not care about the present or the past. It is the observation of all possible things. Although prevision comes gradually, it is the knowledge of what is about to happen.

— Leonardo da Vinci

What will the future of the intelligent era be like in the end? A thousand people have a thousand kinds of looks in their hearts.

At the 2020 Smart China Expo Online, the black technologies displayed in the exhibition hall and the smart life scenes built in the Lijia Smart Park have shown the upcoming smart life to the fullest.

Scenarios like 5G remote driving, smart communities, unmanned restaurants and future colleges are within reach, and people's lives have taken a "smart" leap.

Cultural Tourism Joins Digitalization, and a Smartphone Can Bring You Everywhere

If many of the innovations in smart communities are "expected" in people's daily lives, many practices of smart cultural tourism can be said to be unexpected surprises to the public.

After all, the experience of cultural tourism is difficult to quantify, and also involves complex aspects of transportation, accommodation, excursions, food and beverage. How to balance these different aspects to come up with an optimal solution is a challenge, especially in the post-epidemic era.

In terms of intelligent cultural tourism, Tencent is the leader in China.

As early as 2016, Tencent Cultural Tourism made an overall layout of "Internet + Culture + Tourism", from Longmen Grottoes to a smartphone tour of Yunnan, from the Digital Forbidden City to Digital Dunhuang, accumulating hundreds of innovative cases and implementation services around the world.

On May 15, 2020, China Unicom and Tencent jointly released the *2020 China Smart Culture and Tourism 5G Application White Paper* in Beijing, focusing on 5G culture and tourism industry solutions and implementation practices based on technical research and scenario outlook, and comprehensively demonstrating 5G's ability to promote growth, integration, sustainability, innovation and cooperation in the culture and tourism industry.

Tourism in the post-epidemic era is not simply a return to the same kind of tourism practiced as in the past. It needs more new thinking, new dynamics and new models.

In the "Smart City" pavilion at the 2020 Smart China Expo Online, Tencent T-DAY used smart technology and interactive creativity to create a future smart city exclusive to Chongqing by using the unique elements of the four banks of the two rivers and combining the characteristics of Chongqing's landscape. Tencent T-Day has integrated the elements of Chongqing's landscape and smart life scenes to stimulate the public's interest in smart technology and enhance the understanding of smart city.

When the visitors walked through the sparkling entrance of the exhibition hall, they would hear the Chuanjiang chant and travelled to the classic scenes of ancient Chongqing to experience the historical wisdom and humanistic scenery of the city. If people put on VR glasses and told the destination to "Tencent Xiaowei," all the Internet famous attractions such as Liziba monorail through the floor, Wulong sinkhole and geofracture scenic area and Yangtze River cable car would appear. The visitors could fully appreciate the collision of reality and virtuality.

Realizing to travel from the past to the future through a time tunnel, technologies such as big data centers, 5G networks and cloud service resources have constructed a digital network for city operation, with data flowing incessantly and technology spreading to every corner of the city.

Based on the widely-covered domestic social users, Tencent has combined the cloud computing, big data, artificial intelligence, security and other basic capabilities accumulated by Tencent in the course of its own business development to form professional digital service capabilities in sub-divisions such as all-for-one tourism, scenic spots, cultural and historical museums, special towns, theme parks, green areas and wetlands, and conventions, and to promote the penetration of 5G and other information technology applications into various intelligent scenes.

In addition to Tencent, many of Chongqing's local enterprises also showcased their intelligent cultural tourism "black technology" products at the 2020 Smart China Expo Online. A number of enterprises such as UnionPay, Chongqing Tourism Cloud and Chongqing Travel Square displayed many of the emerging cultural tourism technology applications for the first time, opening up a rich experience from the "cloud" to the "end" for the audience.

The Dazu Rock Carvings World Cultural Heritage Monitoring and Early Warning System, developed and launched by the Dazu Rock Carvings Research Institute, has provided valuable "Chinese experience" for the protection of stone cultural relics worldwide.

From the ancient panda restoration project of Chongqing Nature Museum, which reveals the past life of giant pandas, to the restoration of 3D ancient relics of Chongqing China Three Gorges Museum and the development of the "Mountain and Flowing

Water" Zither WeChat Mini Program based on a thousand-year-old zither "Song Shi Jian Yi," the new integration of digital intelligence, culture and tourism is gradually approaching the lives of ordinary people.

In the Post-Epidemic Era, Technology Enriches Life

In the post-epidemic era, body temperature measurement has become a necessary part of entering public places, especially crowded places such as shopping malls, hospitals, airports and subway stations.

China has, in a short period of time, rapidly established a comprehensive set of precautionary measures through AI technology with the integrated application of rapid body temperature measurement, health codes and trip codes.

For airports, prevention is more difficult because passengers travel across a wider range of regions. Especially for airports with international routes, it is necessary to prevent the disease through a variety of measures. In addition to strict measures such as trip traceability, nucleic acid testing and targeted isolation, it is also necessary to minimize the exposure of people within the airport scenario.

At Chongqing Jiangbei Airport, the construction of "Smart Airport 2.0" is being accelerated. With the continuous development of smart services such as the check-in of self-service baggage, paperless flight services and other intelligent services, it is more convenient for passengers to travel by air and avoid centralized queues such as baggage check-in and boarding pass exchange.

On the south side of Island 3J in the T3A terminal, 16 self-service baggage check-in devices are lined up. Passengers only need to scan ID cards and other valid documents to complete the boarding pass printing and baggage check-in in just a minute. After the baggage is checked, manual sorting is no longer required. The use of big data collection and analysis technology can accurately sort the checked baggage to the corresponding flight. Moreover, passengers can also check their luggage information in real time through their mobile phones.

As one of the most advanced sorting systems in China, the overall length of the baggage system line in T3A of Chongqing Jiangbei International Airport reaches 17 kilometers, and the total number of major equipment in the whole system exceeds 3,000 units. It is the first to adopt open-loop RFID identification application technology in China, which greatly improves the identification rate of baggage sorting. After the system was put into use in September 2017, 140,000 pieces of luggage were processed during the National Day, and the final result was excellent.[1]

In the post-epidemic era, telecommuting and online education will be retained and artificial intelligence will continue to have a more far-reaching impact. Relying on artificial intelligence, cloud computing and other technologies to build innovative

---

[1] Li Xiangbo, Xinhua News, *Explore the Sorting Systems of Chongqing Jiangbei Airport*, October 8, 2017.

operating models and promote cost reduction and business growth will become a key consideration for future industries in the post-epidemic era.

Especially in the medical sector, technology is accelerating to promote the intelligent transformation of the industry. In the application scenario of the identification resolution of the medical device industry of Chongqing SWS Hemodialysis Care Co., Ltd, a "patient" was filled with a catheter, and simulated blood, hemodialysis concentrates and waste liquid flew through the catheter. The blood concentration and other clinical data were displayed in real time on the big screen next to him. By mining these data, doctors could achieve personalized dialysis treatment for each patient.

At the 2020 Smart China Expo Online, the team led by Lan Zhangli, a Professor from College of Information Science and Engineering of Chongqing Jiaotong University, brought a health detection smart toilet, which could detect the contents including urine analysis, urine flow rate detection, body fat detection, blood pressure and heart rate. And the above index contents could be used in combination with smart devices like cell phones and tablets to form a platform of collection end, display end and cloud end system.

Right now, the application of different means such as technologies, data and data science to promote the medical processes including medical consultation, medical records diagnosis, medical reimbursement to digitalization and intelligence has become an important direction in the development of medical information technology. In the past, patients had to queue up for registration and medical treatment for several hours. Now, online appointments, automatic triage and electronic medical record construction can be implemented in real-time medical insurance settlement, which has greatly improved the efficiency of medical treatment, reduced waste of resources, and the satisfaction of doctors and patients increases.

With the development of AI, 5G, IoT and other technologies, the integration of healthcare and technology will also become more in-depth, and provide a broader imagination for smart healthcare.

Lijia Smart Park Brings New Insights

As the permanent venue of the Smart China Expo, Chongqing is making its every effort to build a smart city.

Lijia Smart Park in Chongqing Liangjiang New Area is quite popular as the first smart park in China, a centralized carrier to showcase smart life.

Regardless of cold days or hot days, week days or weekends, the park is always inundated with visitors, which has become a hot destination for citizens and tourists with close-to-nature experiences. Data shows that since the opening of the park in

August 2019, it has received more than 500,000 visitors.[1]

The park is planning to build five zones: Liangjiang ACGN (animation, comic, game and novel), Cloud Forest, Geek Community, Lakeside Smart Core and Innovation Center, forming a functional structure layout of "one park and five zones" and building a never-ending smart stage. The world's top technologies and smart applications are located here, allowing people to experience the smart life of the future city in a "one-stop" manner.

From the moment you walk through the park gate, smart life begins. There is a "face recognition smart storage cabinet" at the entrance. You only need to pass face recognition to store items. Your age, facial features and other information will be visible when you stand in front of "face analysis screen" on the right side of the entrance. After walking around the park, you can check the counts of steps, heart rates and energy consumption by smart screen.

This is just the beginning of the smart experience.

Next, a smart cool road full of "black technologies" will be the "savior" in Chongqing's hot summer. There is a smart spray cooling system on both sides of the smart cool road. As long as you enter the spray area, the smart system will recognize everyone's body temperature and spray water mist according to the body temperature. It could keep the body comfortable, and create the effects of "sunshine censer gives birth to violet smoke."

If you do not want to walk, the unmanned self-driving car in the park will become your caring mobility vehicle. The vehicle can automatically drive according to the established route. It can scan the items 20 meters in the front and make a timely response like braking and slowing down to automatically avoid obstacles 5 meters ahead.

The unmanned small noodle machine, unmanned bun machine, unmanned bowl dish machine will provide services for you when you are hungry. What you just need is to scan the code and place an order, then a special snack of Chongqing will be delivered to your table through the smart machinery and equipment in a few minutes.

Outside the 5G exhibition pavilion, robots were lifting their arms to play the beautiful music "My Motherland and Me" on the piano. Driverless cleaning vehicles were roaming around to ensure the cleanliness of the park and would "actively" avoid colliding with people. You could play a game of hockey or soccer with VR games, or even ride a 5G bike to enjoy the city through the big screen.

The venues such as the Art Fun Pavilion, 5G Pavilion, YunShang Experience Center and Yunding Bazaar, application scenes such as Yunshang Flower Sea, Smart Trail, Yunshang Observation Deck and other experience projects such as driverless

---

[1] Zhong Yi & Han Xiao, China News, *Lijia Smart Park in Chongqing Liangjiang New Area Makes Intelligent Life Within Reach*, October 22, 2020.

cars, robot piano playing and VR immersive experience equipment are exploring the possibilities of future life in all aspects.

At present, Lijia Smart Park has displayed the smart city with 50 scenes and 150 kinds of experiences in 3D among the lucid waters and lush mountains to explore the future life in an all-round way. The Smart Park has preserved the beauty of nature, used advanced smart elements, drawn on the outstanding achievements of the Smart China Expo, built a green and smart interactive experience scene, and created an open and shared urban vitality space. In the future, the Smart Park will bring forward-looking concepts and a global perspective to create a display window for "smart manufacturing town" and "smart city."

In the future, with the development of Chongqing and technology iteration, Lijia Smart Park will be applied to more smart scenes so that people's lives will achieve a "smart" leap.

Ideal Shines into Reality, and Intelligent Era Knocks on the Door

When people all over the world greet each other, they use the word "May all your wish come true," which is perhaps the common ideal of all human beings to achieve what they want just by thinking about it.

In the era of intelligent industry, artificial intelligence is making all kinds of wishes come true.

There is no doubt that the future has come. Between the ideal and the reality, there is no rift, only a distance of a street.

In the world right now, the IoT is not yet fully connected to everything in the world, but it has started to be applied experimentally in the local area with good feedbacks, and it is only a matter of time before everything is connected.

Four major innovations including satellites, drones, autonomous driving and augmented reality are connecting all the world's devices through sensors to create a truly connected society of everything.

The number of sensors that are connected to each device has grown exponentially over the past decade, with sensors on cell phones doubling every four years. IDC predicted that by 2024, there would be more than 50 billion IoT-connected devices worldwide, 11.4 times the amount of mobile phone connections[1].

From the post and telecommunications era to the telecommunications era, from the Internet era to the mobile Internet era, what has evolved is not only the connection of medium, but also the connection of thinking.

When we wake up in the morning, we simply put on our augmented reality contact lenses, which record every conversation, every person crossing the street and everything we see throughout the day. From this constantly observed stream of data, we can use the

---

[1]IDC, *China IoT-connected Scale Forecast 2021-2025*.

personal social and preference data collected to train private AI.

All the sensors can achieve the connection and interaction of people and things, things and things through the smart interconnection.

The data generated and used by IoT encompasses different aspects of daily life such as clothing, food, housing and transportation of human society, and is related to different fields such as public administration, manufacturing, medical and health care and transportation technology.

In the future, everyone will use the brain-computer interface, and the world's known information will be completed in the human-computer interaction of independent learning. People can find a favorite field from a large amount of information to explore the unknown. At that time, humans will be able to exchange ideas through direct communication between brains, and even control machines with ideas. The brain-like intelligence helps people to process data and correlation analysis to solve the problem of general scenarios, and finally realize the concept of strong artificial intelligence and general intelligence.

Some researchers predicted that within the next 20 to 30 years, brain-like AI with general-purpose AI that could pass the new Turing Test may emerge. The maturation of brain-like intelligence can help scientists create artificial brains that will catch up with or even surpass biological brains in terms of storage density, yet consume less energy, which directly gives rise to smarter robots, self-driving cars, medical diagnostics and other AI interactive systems.

In short, in the reality of the ubiquitous IoT and trillions of sensors, the era of intelligence has come as expected. In this era, we can all see that life is making a "smart" leap. Even if we are not the creators of artificial intelligence, we should be the participants of the intelligent era, to meet a new world composed of IoT and sensors, which is closely related to us.

Chapter 4 Achievements:
Cutting-Edge Innovations in Technological Intelligence

Section IV

In the Intelligent Financial Transformation, Technology Brings the "Turbo-Charging" Effects

The sole purpose of science is to alleviate the hardships of human existence.

—Berto Brecht

When it comes to the digitalization of finance, the only thing that comes to most people's mind is digital currency, but behind the market, FinTech has been spreading out on a large scale in recent years and coming into people's daily lives.

Since 2000, the rapid development of Internet technology has changed many industries, and the traditional financial industry is no exception. Because the financial industry has the most abundant and best-quality data, the most fertile "soil" and the largest space in the financial industry, information technology can soundly play its role.

Thus, technology and finance are closely integrated and evolved into a new industry.

Contactless Financial Services

It used to be thought that it would take two years to move from contact to contactless financial services, but it only took two months for financial institutions to adapt to the new challenges.

The COVID-19 epidemic has had a significant impact on the global financial system, prompting a reshaping of the global financial system in the midst of turmoil, and a consensus in the industry to create "contactless finance," allowing FinTech to take a big step forward. During the "Two Sessions," "contactless finance" was again upgraded and became part of the new infrastructure.

During the epidemic, people were unable to leave their homes to meet their business needs such as loans, financial management and insurance claims. In response to these difficulties, many banks, licensed consumer finance companies and other traditional financial institutions have accelerated their efforts to move closer to "contactless finance," with technology investment and business transformation in full swing.

Merchants Union Consumer Finance Co. Ltd., jointly established by China

Unicom and China Merchants Bank, chose a purely online development path with high requirements for technology capabilities at its birth. By embracing new technologies, it took the lead in innovating a purely online, lightly operated Internet consumer finance business model, becoming the first company in the industry to go to IOE for the whole system and the first to have the whole system in the cloud.

During the epidemic, Merchants Union Consumer Finance Co. Ltd. made a big splash. This innovative model relies entirely on the Internet model for customer acquisition and operation, with no direct sales staff and no account manager, and successfully implemented three tasks simultaneously. That is the overall business system on the cloud, the core system de-IOE and the inter-provincial relocation of the company's server room.

During the COVID-19 pandemic, Merchants Union Consumer Finance Co. Ltd. deployed 5,000 intelligent robots[1]. It not only relieved the pressure on labor demand for epidemic prevention and control, but also ensured the normal operation. The intelligent robots independently developed by Merchants Union Consumer Finance Co. Ltd. have the characteristics of low cost, multiple scenarios, high output, high performance and easy tracking, and can recognize more than 200 kinds of user intent with an accuracy rate of up to 99%, taking up 95% of the company's customer service and post-loan asset management work, effectively improving service quality and customer experience.

In addition to Merchants Union Consumer Finance Co. Ltd., most other financial institutions have also opened mobile businesses, forming a matrix of mobile apps, WeChat public platform services, social We Media and channel referrals to achieve fully opening-up service channels.

Especially for personal loan business, customers can "access" with one click, submit document information and capital transaction flow to the platform quickly. The platform automatically identifies the information and completes the registration of information quickly. The intelligent risk control model is opened online. The customer can be quickly assessed and audited, and get the signing information in time to complete the online signing. Successful business approvals are received on the same day or the next day.

A process that takes only one or two hours to complete and relieves many people's urgent needs.

The Power of Technology Empowers SMEs

The outbreak of COVID-19 epidemic has challenged the financial industry's ability to provide fast, accurate and "contactless" financial services, and has provided an opportunity for FinTech to take advantage of online and intelligent services.

The epidemic created a huge impact on SMEs, especially in terms of financing and

---

[1] Zhang Mei, People's Daily, *Merchants Union Consumer Finance Co. Ltd. Deployed 5,000 Intelligent Robots*, February 18, 2020.

risk control, and contactless service finance has become a life-saving remedy for SMEs.

For the long-tailed micro and small enterprises with small scale and low credit limits, their capital demand has the characteristics of being short, small, frequent and urgent. At the same time, the enterprises themselves have high elimination rate, irregular financial information and lack of effective asset collateral. Therefore, in the past, it was difficult to get financial support under the traditional credit granting model. After the epidemic, the situation of small and micro enterprises was even more difficult as it made the work of offline bank branches and credit staff more difficult.

In order to solve the problem of difficult and expensive financing for SMEs, Financial One Account is the first in the industry to launch the "Integrated Platform for SME Services (Anti-Epidemic Edition)," which helps the government conduct in-depth analysis of macroeconomic and meso-industry through big data models, predicts the development prospects of the industry, forms enterprise portraits and screens the whitelist of epidemic companies. Through the four tools of intelligent planning, intelligent services, intelligent products and intelligent operation, Financial One Account could build SME service platform and anti-epidemic zone to help SMEs resume work and production.

The platform also realizes one-click matching application of financing and subsidy policies through Askbob intelligent policy search and recommendation engine, which has integrated more than 6,000 policies and can quickly match relevant policies for enterprises through AI semantic technology.

With new technologies such as cloud computing and big data, FinTech is able to serve customers online. During the transfer from offline to online activities triggered by the epidemic, FinTech platforms can streamline traditional processes, reduce intermediary links, reach long-tail customers without contact, reduce the resistance caused by information asymmetry, and provide more efficient and convenient services to the customers.

Overall, with the joint efforts of many parties, the financing for small and micro enterprises in 2020 achieved remarkable effects of "volume increase, price decrease and business expansion." By the end of 2020, the balance of inclusive micro and small loans was 15.1 trillion yuan, an increase of 30.3% year-on-year, and a total of 32.28 million micro and small business entities were supported throughout the year, an increase of 19.4% year-on-year[1].

FinTech, the new product, has driven many changes in the financial industry, optimized the structural reform of the financial supply side, helped the digital transformation of the banking industry, empowered the transformation and development of regional banks' risk control, established a new financial system of openness and

---

[1] Yao Junfang, Xinhua News, *The Balance of Inclusive Micro and Small Loans Exceeded 15 Trillion Yuan! What Does the Rapid Growth of Micro and Small Loans Depend on?*, January 22, 2021.

sharing, and helped financial institutions achieve "three rises and two drops," that is, the improvement of revenue, efficiency and service quality, and the reduction of risks and costs.

China's FinTech Development Is Leading the World

In the future, will finance dominate technology or technology dominate finance?

In the case of China, the wind of technology was clearly stronger than finance in the early years. BAT and other technology giants competed to enter the financial sector first, and a large number of P2P, crowdfunding and online lending companies were born, which brought huge pressure to traditional financial institutions. In the past two years, traditional financial institutions have gradually started to vigorously promote the combination of technology and finance themselves. They have set up technology companies, paying more attention to the role of technology in promoting finance.

The engine of a car, through turbocharging, can increase the air intake, thus increasing the power and torque of the engine and making the car more powerful and fuel-efficient. Financial institutions can improve efficiency, reduce costs, provide better user experience and make finance better by "turbo-charging" with technology.

With the help of AI technology, the traditional banks are possible to achieve intelligent customer service, intelligent identity authentication, intelligent operation and maintenance, intelligent investment advisory, intelligent claims, anti-fraud and intelligent risk control. The application of big data technology enables banks to collect information from various channels more widely for analysis and risk management, and to conduct accurate marketing and customer acquisition. The application of mobile Internet technology greatly expands the market space of banks and allows people to conduct financial transactions anytime and anywhere.

Obviously, finance and technology are both intertwined, but each in its own place promotes its strengths. Currently, this is the most accepted integration of all parties. As to most traditional financial institutions, the transformation catalyzed by the COVID-19 epidemic is a "self-revolution." "Contactless Finance" is not just about transferring of inherent business to online. It essentially means a linked change in business philosophy, business model and personnel structure.

On December 17, 2020, the *China Financial Technology and Digital Financial Inclusive Development Report (2020)* released by the Zhongguancun Internet Finance Research Institute shows that China's financial technology industry is in the forefront of the world, and the revenue of financial technology in 2019 is about 1.4 trillion yuan. In the same year, China's financial technology financing accounted for 52.7% of the world's total[1].

---

[1]Pan Fuda, Beijing Daily, *China's Financial Technology Industry Is in the Forefront of the World*, December 18, 2020.

Overall, China's FinTech innovation is mainly centered on big data, AI and blockchain. The future of FinTech is likely to drift towards blockchain. FinTech innovation will be dominated by large institutions such as major ICBC, BCCB and WeBank.

There is no end to development and innovation. Finance is the bonding agent of the market circulation and social operation. After being integrated with the new generation of information technology on a large scale, it will feed back the technology industry and help the research and development of new technology and business innovation, so that the intelligent era will come earlier.

Section V
The Convenience of Life Starts from the Smart Government Services

To live in the world is to have the obligation to make it better.

— José Martí

From the beginning of people's expectation of the intelligent era, smart government services have been the nodes to be pre-driven.

Smart government services such as hotlines, administrative halls and online offices are bound to move to more important positions, which are not only the most critical positions to face the challenges of people's livelihood and services, but also play the role of an important route to the future intelligent development of China's government services data.

There is a direct cause-and-effect relationship between smart government services and smart cities.

Streets, street lights, buildings, weather and livelihood policies are all within the range of the smart government services.

Chongqing's "Answer Sheet" to Smart Government Services

Since Chongqing was listed as a national smart city pilot in 2013, smart home, IoT, autonomous driving, 5G and other technologies have taken roots in Chongqing, and the Smart China Expo has been settled here for a long time. Chongqing is developing rapidly along the road of intelligence.

In terms of the smart government services, Chongqing began to layout early.

In November 2014, Inspur Group established a cloud computing center in Chongqing Liangjiang New Area from start-up to operation, and it has been 7 years since then. In these years, Chongqing Government Services Cloud has delivered its "answer sheet" to hardware, software, and then to service.

As of January, 2021, the number of Chongqing's "Cloud Chief" units has reached 110, promoting 2,458 information systems to the cloud. The rate of cloud has risen from 26.6% to 98.9%. The government services cloud "a cloud bearing" service system has formed through the "Digital Chongqing" cloud platform[1].

On the huge display at the Smart City Operation and Management Center in Jiangbei District of Chongqing, data scroll in real time, monitoring is online in real time, and the operation status of the district-wide urban management system is clear at a glance.

The smart city management system can supervise and manage the city facilities and environmental sanitation in an all-weather and full-coverage manner. The system is connected with 25,000 video monitoring cameras, covering all key road sections in the region.

If it rains and water gathers on the road, the video capture system will immediately capture and prompt the duty officer, who will review the information and notify the relevant departments to deal with it, and it only takes ten minutes for the road to return to normal.

If the manhole cover is tilted, the sensor will automatically warn, so there is no need to worry about the falling of passers-by. If the street lights are not on, the maintenance personnel will determine the location through the current change and rush to the scene immediately. These small problems, which are unavoidably overlooked by manual investigation, have nowhere to hide in the system, and the disposal is more efficient. By the end of July 2020, Jiangbei District handled a total of 2.49 million urban management cases with the 92.33% of public satisfaction rate[2].

Apart from Jiangbei District, the High-tech Zone Government Services Center Office Center also introduced "smart government service system + smart government services self-service terminals" built by the Inspur Group to eliminate citizen's time wasted in between. With "one network" online and "one window" offline, all businesses can be completed at one time, making things more convenient and efficient for the general public.

At the same time, the high-tech zone also has launched online "Scientific Government Services" and opened a "guide" section. The function can provide enterprises and the general public online query office guide and online reporting services. Creating online "One Network" service so as to achieve a unified entrance, unified navigation, unified authentication, unified declaration, unified query, unified interaction and unified evaluation.

In addition, Inspur Group also assisted in the development of the "Scientific Government Services in Wechat Office Hall," also called "Portable Version of the Office Service Hall." Through WeChat, it can realize the functions of reservation, online declaration, consultation feedback, complaint and evaluation, which makes

---

[1]Xiayuan, Chongqing Daily, *The Construction of Smart Chongqing Is Accelerating, and 18,500 Digital Economy Enterprises Have Gathered in Chongqing*, January 21, 2021.
[2]Cui Jia & Liu Xinwu, People's Daily, *Intelligence Makes Cities Better*, August 29, 2020.

the government services more diversified and makes the mobile government more convenient for the public.

In December 2020, Chongqing's collaborative office cloud platform "Yu Kuai Zheng" was put into operation. "Yu kuai Zheng" is an integrated, intelligent and digital government services platform built by Chongqing, which has a city-wide government directory, basic government collaboration functions such as integrated communication, audio and video conferencing, city-wide official document exchange, and more than 40 general office functions such as meeting management, meeting collaboration, document flow, inspection and supervision, and affairs management.

It can realize "cross-level, cross-department, cross-region, cross-business and cross-system" collaboration among parties, government organs, villages, community organizations, enterprises and institutions in the city, and improve the efficiency of government services.

Relying on "Internet +," cloud computing and big data, all districts and counties in Chongqing have completed the construction of digital city management platform. Intelligence not only makes the city better, but also becomes a personal experience for more and more citizens.

Singapore E-Government Services Bring Inspirations

Since 1980, Singapore has put forward projects like the "National Computerization Plan," "National IT Plan," "IT2000 — Intelligent Island Plan," "Information and Communications Technology 21st Century Plan," "Intelligent Nation 2015" (iN2015) and "Smart Nation 2025" and began to explore in the field of smart government services early.

Unlike Chongqing, China, which is centered on the government services cloud platform and has access to all regional government services systems, Singapore is centered on SingPass account. Through the SingPass platform, one account can access all government-related systems and departments, including CPF provident fund system, IRAS tax system, ICA Immigration Office, MOE Ministry of Education, LTA Ministry of Transport, Property Management and Ministry of Labor. Almost all government services can be included.

All websites of government services in Singapore cover almost everything that residents need in their daily lives, providing them with "services from birth to death," which has also contributed to the development of Singapore.

Singapore is a leader in building e-government services. Since June, 2020, some Singapore government departments had installed cameras. Since then, people could enjoy the services provided by government departments just by face recognition rather than ID cards. The system is interoperable with SingPass Mobile, which was launched in 2018, and allows people to use SingPass Mobile to register their fingerprints and face information in the government's biometric database, and later scans their faces with a camera for comparison.

In addition, Singapore has developed the Corpass Enterprise Management Platform, where companies can complete government services in a one-stop platform, including applications for subsidies, release of recruitment, tax payment, provident funds, change of company information, application and cancellation of work permits. It is similar to an amalgamation of the Industrial and Commercial Office, Social Security Office and Tax Office.

In 2020, the department of Singapore government services had created real-time virtual VR training platform HTS2, which is the world's first training platform that combines real-time and virtual elements. Featuring the immersive nature, the system allows the trained team members to use their own workstations in a physical environment. And then, this system can also simulate the real-life environment in a highly realistic environment, in which the restored buildings including those landmarks are all lifelike.

Currently, the HTS2 system has been deployed in four government departments, namely Singapore Police Force, Singapore Civil Defence Force, Immigration and Checkpoints Authority and Singapore Prison Service, with over 1,000 police officers participating in the training.

Singapore's e-government brings inspirations to the field of smart government services in terms of inclusiveness and openness, data transparency and data interconnection, which makes government services easier and more convenient.

Building the Smart Government Services Is the Inevitable Future

Guided by a new generation of information technology, government services have a whole new set of characteristics, including personalization, cloud orientation, data orientation and intelligence. Government operations and government services are more targeted and efficient, and smarter in understanding the needs of the public, thus opening up a new era of smart government services.

Compared with previous government services, smart government services emphasize social services, pay more attention to the needs of individual members of the public, and are smarter and more predictable.

The government not only has big data resources that ordinary enterprises and individuals cannot have, but also has the administrative power to obtain all kinds of necessary external social data.

In terms of data owned by the government itself, it includes data on industry, commerce, taxation, public security, transportation, medical, health, employment, social security, geography, culture, education, science, technology, environment, finance, statistics and meteorology. In terms of external data, the government has the ability to obtain social data, such as world economic operation data, public opinion data and enterprise report data.

Big data applications in government services have changed the traditional problems of hindsight caused by the government not being on the front line, and can be used to predict and warn social and economic problems in advance through big data modeling to achieve more effective governance and social services.

In the context of the new round of technological revolution and industrial changes, the "three in one" integration of government cloud, big data and artificial intelligence has become the main direction of smart government services. For example, the government services cloud is currently the most mature area where new technologies are combined with government services, including a large number of application examples from both the central government and local governments. The health code, which was put to use during the epidemic, has now become a must-have for Chinese people to go anywhere else.

Smart government services have the foundation and infinite possibilities for the future, and we are ushering in a new era of government services.

Chapter 5

Chongqing: A Key Witness of the Intelligence Era

A city will never lack its own culture and charm, nor will it lack its own business and vitality. The former is cultivated in the history of the city, and the latter comes with the mobility of people. However, the fates of one city and another are often quite different. The vitality of an ordinary city mainly depends on how many lively scenes there are on the ground; the vitality of a great city mainly depends on how gorgeous blueprints there are on the horizon.

Chongqing, in the global smart industry, may not be able to dominate the smart industry. However, taking the past several years as the observation coordinates, it is always exciting and amazing. With the construction of strategies from imagination, and the touch of the future with practices, Chongqing's determination to embrace the intelligent industry, thoroughly and hard, is enough to become an example for many cities.

Chongqing is a watchtower for observing the development of the global intelligent industry, and a navigator who redefines itself by taking advantage of the opportunities of the times.

Section I

A City of Imagination: Chongqing Is Committed to the New Trend of Intelligent Industry

The world of reality has its limits; the world of imagination is boundless.

— Jean-Jacques Rousseau

On this blue planet where we live, humans have identified more than 1.7 million biological species, and the actual number is much higher than that. But why is it that humans can surpass all species and build a radiant civilization on the Earth?

Numerous historians have given completely different answers to this question from different perspectives, and the new and prominent historian Yuval Heraly, in his book *A Brief History of Mankind: From Animals to God*, gives a completely new answer: Because humans have the ability to create and believe in fictional things and stories, that is, the ability of human imagination.

Gods, nations, peoples, corporations, money... all the constituent elements of modern civilization that have become the basic consensus have a common source of shared human imagination.

Human beings live in groups, so there is a city, and the imaginations of the citizens gathered in the city become the imagination of the city. Of course, different cities have completely different imaginative abilities.

Chongqing is located in the western part of the country and has no advantage of the basic environment of intelligent industry. However, it was first listed as a national smart city pilot, and then has become a permanent venue for the Smart China Expo. The city is given a new vitality in the intelligent era.

Chongqing is trying to make itself a city of imagination in the era of intelligent industry.

IoT in Chongqing Blossoms

"Chongqing becomes a national smart city pilot!" This is a new business card which was achieved by Chongqing several years ago.

What exactly is a smart city like? Looking back at the news reports of the year, people relied on their own imagination to build and define the smart Chongqing in their minds. Some companies have developed mobile clients to display various information of the city and divided it into six sections, including wonderful city, ecological city, special industries, government services, investment and famous enterprises, thinking that this is the smart city. Some people have developed a set of virtual clothes fitting "magic dressing mirror," thinking that this is also a smart city.

For a city, eight years is just a snap of the fingers. But now, Chongqing's layout in the smart industry has long surpassed what was imagined at the time.

As early as December 2010, China's the Ministry of Industry and Information Technology has officially approved Chongqing Nan'an District to be the "National New Industrialization Industry Demonstration Base of Electronic Information (Internet of Things)." It is also China's first national IoT industry demonstration base. Thanks to Chongqing's position as a national smart city pilot, a number of domestic and foreign IoT enterprises have settled in the Nan'an IoT Industrial Park, and the IoT industry clustering effects gradually emerged.

At present, the industrial park has formed a pattern of coordinated development of large, medium, small and micro enterprises in the industrial chain based on the leading industries, supported by a large-scale IoT operation platform. A total of 21 enterprise technology centers or R&D centers above the provincial level have settled in the park[1], integrating IoT R&D, testing, operation and service platforms, forming a full range of IoT platforms.

The base has promoted the development of a number of large-scale operation platforms and public service platforms such as China Mobile IoT and OneNET IoT open cloud platform through open sharing, product services and other development methods in transportation, environmental protection and other industrial fields. At present, the base takes networking of vehicles and industrial IoT construction as industrial contents, and is gradually forming a complete industrial system integrating IoT chips, modules, terminals, platforms and applications to achieve technological integration and innovations in the IoT industry chain.

At the end of 2014, the Nan'an Cha Yuan was still full of construction sites, and now the once proud smart home is just taking shape. But at that time, a dozen smart home companies gathered in the IoT Industrial Park have started brainstorming. Since data interaction is difficult to land, then is it feasible to connect electricity? Since the smart ecology is too far away, then is it possible to develop the operating system first?

With the construction of Chongqing's public cloud and government cloud, the data interaction in the conception of the smart home was realized, accordingly, the smart home also moved from simple remote control in the early stage to semi-intelligent control for whole house, and then integrated with 3D sensors, IoT and other technologies to build today's smart home.

---

[1] Rao Jinlan, Nan'an.com, *The Nan'an IoT Industry Demonstration Base Has Been Awarded the Five-Star National Title for Two Consecutive Years*, May 8, 2020.

While tackling the core technology of smart home, Chongqing has also started to lay out the industrial chain of smart home manufacturing.

In the production park of Huida Smart Home (Chongqing) Co., Ltd, an Industry 4.0 smart sanitary ware production line has been put into operation. The whole process of product production adopts automatic logistics to convey equipment. With the whole production process online, automatic 3D storage among each process is set up, and the product varieties are automatically identified through RFID barcode management system. The products are automatically delivered according to orders, making the production more flexible.

As to the future imagination of smart home, the under-construction Chongqing Luo Huang Smart Home Town is wielding inspiration. In the future, the town will become a smart home standard-setter and trend-publisher. Chongqing has become one of the key areas of the national IoT industry. From the real-time perception of quality changes of the Yangtze River and Jialing River water to smart monitoring of traffic conditions, from significantly improving the efficiency of enterprise production to promoting the innovation of industry chain reconstruction, more and more IoT applications in Chongqing have taken roots and blossomed.

At the 2020 Chongqing IoT Integration Development Forum and Members' Conference, the top ten application cases of IoT in Chongqing in 2020 were recommended. A large number of typical demonstrations emerged in the fields of smart transportation, smart medical care and smart security, all of which have been tested successfully in Chongqing, with reproducible business models and strong promotion value.[1]

Evolution Under the Industrial Agglomeration Effects

At present, Chongqing has gathered the four trump cards of the intelligent era of IoT, AI, big data and 5G. The door of the intelligent era has been slowly opened.

The imagination space of the intelligent era, with the support of the four trump cards, is still continuing to spread.

In terms of IoT, Chongqing has not only formed a complete industrial chain, but also achieved ten mature successful cases of implementation applications.

In terms of artificial intelligence, Chongqing has built a national new-generation AI innovation development experimental zone which has made breakthroughs in a number of key core technologies and achieved initial results. For example, the local enterprise, Chongqing Xin'anbijie IoT Technology Co. has created the intelligent safe infusion control system. The system is divided into four modules — drip rate infusion controller, nurse station management center, smart wearable devices, and Zig Bee wireless network. The smart wearable devices worn by nursing staff can push the fluid variation of the patients under their supervision in time, so that the nursing staff can

---

[1] Han Menglin, Xinhua News, *The Top Ten Application Cases of IoT in Chongqing in 2020 Are Recommended*, December 25, 2020.

know the infusion status of the patients in time, no matter where they are.

Relying on the achievements of IoT, AI, big data and 5G, Chongqing has implemented the "Cluster Development of Strategic Emerging Industry" project to rapidly expand the scale of emerging industries. Chongqing focuses on the new round of technological revolution and strategic trends in industrial transformation and accelerates the cultivation and construction of a number of strategic emerging industrial clusters. Chongqing also promotes the clustering, integrated and ecological development of the new generation of information technology, alternative fuel vehicle, intelligent connected vehicle, high-end equipment, new materials, biomedicine, energy saving and environmental protection, and software information services.

During the "14th Five-Year Plan" period, Chongqing's total industrial volume exceeded 3 trillion yuan, and the industrial added value increased by 6% annually on average, accounting for 30% of GDP, and a batch of tens of billions of industrial clusters was built[1].

Under the effects of industrial agglomeration, "industrial integration" has become a high-frequency concept in the intelligent industry, especially the in-depth implementation of smart manufacturing, which promotes the in-depth integration of a new generation of information technology and manufacturing, intensifies the upgrade of enterprise equipment and technological transformation, and promotes industrial development. The high-end growth of the value chain has made Chongqing's overall level of integration of informatization and industrialization stand at the forefront of the country, unleashing the charm of the smart industry.

At the same time, fields such as autonomous driving, smart medical care and smart governance services are all undergoing rapid upgrades under the effects of industrial agglomeration.

Currently, autonomous driving has been integrated into 5G, resulting in two major directions, 5G remote driving systems and autonomous vehicles. Smart medical care has evolved from surgical robots to 5G remote surgery and go hand in hand with Internet hospitals. Smart government services not only cover the whole area, but also go deeper in urban transportation and livelihood services, and open up the data between various departments, unlocking the functions of "one-click processing."

The New Trend of the Intelligent Industry

There is no doubt that Chongqing has become the new highland of China's intelligent industry.

The most cutting-edge technology is born here, the most up-to-date ideas collide here, and future ideas are unfolded here.

Here, everything one can imagine about the intelligent industry is satisfied.

---

[1]Zheng Sanbo, Chongqing Morning Paper, Chongqing's Total Industrial Volume Exceeded 3 Trillion Yuan, February 19, 2021.

Since 2017, Chongqing has started an intelligent transformation in the manufacturing industry, using big data smart technology to transform and upgrade the traditional manufacturing industry, and vigorously cultivate intelligent industrial chains such as big data, artificial intelligence and smart hardware. By the end of July 2020, Chongqing has completed a total of 2,200 intelligent transformation projects, built 67 digital smart factories and 539 digital workshops.[1]

In addition, Chongqing City is striving to build the whole industry chain of "chips, screens, smart terminals, core elements and network," to seize the opportunity of the new round of development of IoT, and to give full play to the platform operation and application demonstration advantages. Chongqing is also striving to make up for the shortcomings of MEMS sensors, communication module design and manufacturing to build an industrial system of "three in one" of hardware manufacturing, operation services, and system integration.

Chongqing's smart industry has released the signals of industrial layout, which has sparked widespread interest. Having noticed the broad prospects of the western region, more enterprises are willing to settle in Chongqing.

At the same time, a new trend of intelligent industry of developing the southwest China is also rumbling forward in a wider scope.

In recent years, the pace of intelligence in Sichuan, Guizhou and other western provinces is also accelerating. Sichuan Province has introduced *An Implementation Plan for the Development of New Generation of Artificial Intelligence in Sichuan* and other action plans. The development index of big data industry in Guizhou ranked third in the country and it has been ranked in the forefront of the country for many years.

The western region of China, with 400 million people, accounting for two-thirds of the country's land area, has huge development potential, obvious cost advantages of production factors, where intelligent industries are on the rise.

---

[1] Zhao Yufei & Wu Kunpeng, Xinhua News, *Western China Is Emerging As a New Highland of Intelligent Industry,* September 14, 2020.

Section II

A City of Strategy: From a Manufacturing Town to a Smart City

Those who set great ambitions not only have superb talent, but also have the will of perseverance.

— Shu Shi

Sun Tzu's *Art of War* says, "Tao is the spirit of art, and art is the body of Tao; unify art with Tao, and get Tao with art."

In Chongqing, the "Tao" is the city's strategy, and the "art" is the intelligent industry.

From "manufacturing town" to "smart manufacturing town," Chongqing has started the transformation process for less than ten years. The whole process is clear and assertive, from IoT base to big data center, from data interaction to government on the cloud, from the famous enterprises landing to the growth of everything. Finally, with the shortest time, Chongqing has become the new highland of China's intelligent industry.

All these depend on the right "Tao."

Strategic Evolution of Smart Cities

What kind of road should a "perceptive, breathing and warm" city of the future choose?

Before Chongqing, although there are cities around the world experimenting with the path of smart cities, all of them are still unformed and still in constant exploration. Chongqing is no different, and it is also groping to correct its path.

In June, 2018, Chongqing released the *Action Plan for Chongqing Municipality's Innovation-Driven Development Strategy Led by Big Data Intelligence (2018-2020)* at the first Smart China Expo to implement the big data intelligence development strategy.

The strategy proposes to accelerate the digital industrialization and industrial digitalization, focusing on intelligent connected vehicles, smart manufacturing, smart perception, smart IoT, smart robotics, smart terminals, integrated circuits, cloud computing big data (supercomputers), human-computer interaction and other industrial fields. An open and collaborative big data intelligent industrial innovation system has been initially established.

To complete this big plan, 5G is indispensable.

In 2019, Chongqing released *Chongqing Action Plan for Accelerating the Development of 5G (2019-2022)* to seize the opportunity of 5G development, and accelerate the deployment of 5G network, industrial development and commercialization in Chongqing. The action also focused on promoting the construction of "smart manufacturing city" and "smart city," and cultivating new dynamic energy for economic growth.

In the same year, Chongqing also released the *Chongqing New Smart City Construction Program (2019-2022)*. The program clearly states that a first-level national demonstration city of big data intelligent application, urban and rural integration and development of smart society model will be built by 2022, and the innovative construction of a new smart city will drive Chongqing to achieve a new round of leapfrog development.

In the successively issued strategic plans and action plans, Chongqing's route for the transformation and development of smart industries is clear, and each step indicates Chongqing's transformation from manufacturing city to the "smart manufacturing town" and "smart city."

The Rebirth for Old Industrial Cities

When it comes to China's smart manufacturing, people may first think of the first-tier cities like, Beijing, Shanghai, Guangzhou and Shenzhen, as well as the eastern coastal provinces. Chongqing, located in the southwest, seems to be difficult to be reflected in the mind, which is actually a cognitive misunderstanding. Chongqing, as a city with early industrial development in modern China and one of the important old industrial bases, has long been a step ahead in modern manufacturing upgrading, forming the world's largest electronic information industry cluster and China's leading automobile industry cluster, becoming the world's largest laptop computer production base and the world's second largest cellphone production base. In the fields of equipment manufacturing, integrated chemical, materials, energy and consumer goods manufacturing, Chongqing also has its own industrial advantages, and the scale of industrial clusters is also hundreds of billions.

The old industrial cities are ready to develop smart manufacturing.

Chongqing is located in the upper reaches of the Yangtze River, and the unique city culture formed over the centuries, where inclusiveness and openness naturally exist. In this urban cultural atmosphere, Chongqing people's character also breeds a kind of "if you persist, anything could be possible" jianghu[1] gene. The transformation from an inland trade capital to a heavy industrial city, and now from an industrial city to a smart city, the wharf culture and the jianghu gene also provide a constant source of energy

---

[1]Jianghu, in Jingyong's martial arts works, metaphorically refers to an alluvial underworld of hucksters and heroes beyond the reach of the imperial government.

behind the grand narrative of the high level.

As the connecting node of "Belt and Road" and the Yangtze River Economic Belt, and the inland international logistics hub, Chongqing is taking a new attitude to meet the opportunities and challenges of the new era.

In the planning of smart city, smart manufacturing has been the key point. Major local manufacturing enterprises also come on stage in this roaring wave of smart manufacturing.

Zongshen Group, a leading motorcycle manufacturer in Chongqing, has always been the vanguard of intelligent transformation. Its Industrial Internet full-industry chain innovation platform, Humi.com, has built Humi Yunxi, which is the first Industrial Internet identification resolution secondary node platform in the motorcycle industry in Chongqing and even the first one in China. As of December 3, 2020, the total number of identification registration has exceeded 4 million, the total number of identification resolution has exceeded 40 million, and the average daily resolution volume has reached nearly 300,000, which helped Chongqing enter the seventh place of national cities in terms of identification resolution.[1]

At the 2020 Smart China Expo Online, Humi.com brought a mini version of "Zongshen 1011 Intelligent Production Line" to the site, allowing the global audience to see the production process of the entire assembly line. Through the application of marking, after the Zongshen 1011 production line has been intelligently upgraded, the overall efficiency of the Zongshen motorcycle assembly production line has increased 4 times, the personnel has been reduced by 70%, the automatic error correction and prevention capacity has increased 10.6 times, and the operation automation rate has increased 10 times. The overall production of Zongshen Group has achieved a 25% increase in equipment utilization rate, 15% increase in production efficiency, 20% reduction in defective products, and 20% increase in equipment health.[2]

Chongqing's automakers are also moving closer to intelligence, using the Industrial Internet as the core and transforming their own production plants while helping other manufacturers in the industry chain to carry out smart production.

Chang'an Automobile is one of the pioneers. Chang'an Automobile, together with Chongqing Unicom, has forged a collaborative smart manufacturing factory based on "5G + Industrial Internet" and built a global network collaborative design and manufacturing system, connecting tens of thousands of enterprises in the industrial chain, value chain and resource chain. In the smart manufacturing factory, the whole production elements, such as people, machines, materials, methods and environment, rely on 5G connection. 5G technology carries out real-time monitoring

---

[1] Yang Ye, Shangyou News, *The Core of Industrial Internet Identification Resolution Is to Empower Enterprises*, December 3, 2020.

[2] Zheng Sanbo & Peng Chen, Shangyou News, *Exploration: What Is the Experience of Bringing the Mini Version of the Zongshen 1011 Smart Production Line to the Site?*, September 14, 2020.

of the production operation status in the factory floor, including personnel, equipment, production capacity, energy consumption and logistics so as to achieve transparent factory management and unmanned production.

With the support of smart innovation, manufacturing enterprises have been given a new lease of life, and Chongqing has gained new growth momentum, attracting industry giants to settle in Chongqing.

In October 2020, Geely's Industrial Internet global headquarters was officially located in Chongqing. Based on Geely's thirty-five years of manufacturing experience and resources, Geely's Industrial Internet Platform has three core competencies of strong manufacturing heritage, multiple application scenarios and full coverage areas, and provides platform-level services for industrial transformation with cross-industry, cross-field and multi-scene solutions. By building a digital base integrating resource energy efficiency, security and trustworthiness, data intelligence and intelligent IoT, the Geely Industrial Internet Platform enables enterprises to have integrated basic capabilities for digital transformation, supporting five solutions of factory digitalization, digital operation, C2M flexible customization, smart travel and dual carbon management, promoting high-quality industrial development and creating high-quality life for society.

The City of the Future Is on Its Way

The future is here. As a smart city takes shape, Chongqing is already on its way and is about to see the dawn.

In the new year, Chongqing also has a new plan. In the intelligent transformation and upgrading planning of 2021, Chongqing Municipality has drawn up the goal of promoting the implementation of 1,250 intelligent transformation projects[1], including the implementation of digital equipment, digital workshops, smart factories, Industrial Internet platform construction of "Cloud on the Platform" and the application of new models of smart manufacturing projects.[2]

Intelligent transformation is conducive to improving the level of industrial enterprises, intelligent transformation and upgrading of the manufacturing industry. It has a vital role in reducing costs and increasing efficiency, iterative upgrading of the products and achieving high-quality industrial development.

In the fields of 5G base station construction, ultra-high voltage, intercity high-speed railroad and urban rail transportation, big data center, artificial intelligence and Industrial Internet, Chongqing is boosting innovation by building a solid "hardware foundation" in order to seize new opportunities, expand new space, cultivate new dynamic energy, and equip "wings" for Chongqing's economic development.

---

[1]Chongqing Economic and Information Commission, *Notice on Carrying Out the Identification Work of Chongqing Intelligent Reconstruction Project in 2021*, February 23, 2021.

[2]Liu Hanshu, Shangyou News & Chongqing Morning Paper, *Decomposition Criteria of Chongqing's 1,250 Intelligent Reconstruction Project*, February 23, 2021.

At present, Chongqing has built 42,000 5G base stations and achieved full coverage of 5G networks in the key areas of all districts and counties in the city.[1]

In order to strongly support the construction of new infrastructure such as 5G, Chongqing Municipal Communications Administration released the *Chongqing Municipal Land Spatial Planning Communication Professional Planning-5G Special Planning* in 2020. During 2020–2025, Chongqing will build an ultra-high-speed, high-capacity, intelligent, ubiquitous and perceptive communication infrastructure around the 5G communication network, in accordance with the overall spatial layout of the city's "one district and two clusters" and in conjunction with industrial development, to achieve "one plan, one construction and one development" by the end of the planning period, and build 150,000 5G base stations across the city to achieve the goal of global leadership in the overall level of 5G services.[2]

With strategic planning that is carried forward gradually and intensive practices of intelligent transformation, Chongqing today is already taking the shape of the city of the future.

The future is waiting for us to knock on the door.

---

[1] Huang Guanghong, Chongqing Daily, *Chongqing Has Built 42,000 5G Base Stations and Achieved Full Coverage of 5G Networks in Key Areas of All Districts and Counties in the City*, July 3, 2020.

[2] Li Shu, Shangyou News & Chongqing Morning Paper, *Chongqing Will Build 150,000 5G Base Stations by 2025, According to a New 5G Infrastructure Plan*, May 15, 2020.

Section III

A City of Practice: From the Landing of Famous Enterprises to Driving the Rapid Development of Various Industries

The best time to plant a tree was ten years ago, and the second-best time is now.

— Dambisa Moyo

Eleven years ago, when Chongqing built its headquarters city, it had no idea that this tree could grow so luxuriantly.

Headquarters of several Fortune 500 companies have settled in Chongqing one after another, bringing cutting-edge technologies, new trendy ideas, and valuable experiences derived from continuous trial and error.

The big tree also brings a green shade. Standing on the shoulders of giants, local enterprises continue to innovate and drive into the fast lane of development.

Famous Enterprises Take Root and Bring Cluster Effects

Ten years ago, when China was surfing in the blue ocean of Mobile Internet, "artificial intelligence" was still a relatively unfamiliar term. Within a decade, China's AI industry has entered a phase of explosive growth, driven by national policies and corporate innovation.

In this process, the cluster effects brought about by the landing of famous enterprises, become bigger like a rolling snowball, bringing new opportunities to Chongqing.

Now, Alibaba, Tencent, Intel, Xiaomi, Baidu, Geely and other companies have settled in Chongqing, and have launched in-depth cooperation with Chongqing in many fields, such as production, learning, research and application, to achieve mutual benefits and help the intelligent innovation of Chongqing develop rapidly.

Since 2018, Chongqing has successfully held the Smart China Expo for three consecutive years, drawing on the wisdom of global smart industry development and bringing together national smart industry resources to land. One step at a time, Chongqing has seamlessly climbed up the ladder, showing a strong confidence for the outside world to transform and accelerate its development.

During these years, Chongqing has initiated in-depth cooperation with many leading digital economy companies, including Tencent, Huawei, Alibaba and Inspur in numerous fields.

From March to April of 2020, in just one month, eight famous enterprises have settled in Chongqing. The innovation in the smart industry has become the most central driving force.

Automobile industry is one of the important industrial manufacturing categories in Chongqing. Auto famous enterprises began to deploy in Chongqing many years ago. Now it is the time to increase their bargaining chips. At the end of March, 2020, Ford Motor's first domestically-produced Lincoln "Adventurer" under the Lincoln brand rolled off the assembly line in Chongqing. On April 23, 2020, Liangjiang New Area Management Committee, Dongfeng Motor and Xiaokang signed the *Agreement on the Joint Construction of a Medium-and-High-End Alternative Fuel Vehicle Project* to promote the development of the medium-and-high-end alternative fuel vehicle project of Chongqing Jinkang New Energy Automobile Co., building it into a medium-and-high-end alternative fuel vehicle benchmark enterprise with an annual production capacity of 150,000 vehicles and an output value of over RMB 30 billion by 2025.[1]

On April 2, 2020, Fosun Group deepened its strategic cooperation with Chongqing and would participate in some major transportation projects.

Tencent, Alibaba, Inspur, Huawei once again have deepened the cooperation with Chongqing in artificial intelligence, 5G and research institutes. For example, Ali will establish 20 industrial livestreaming bases in Chongqing's well-known industrial belt area and jointly promote the "C2M Super Factory Plan" to create 20 factories with over 100 million in sales[2]. Tencent will take the advantages of Tencent game photon studio group and other traffic advantages to improve the added value of tourism commodities and surrounding agricultural products in Pengshui, and to promote the sales of the agricultural and sideline products, new cultural and creative peripheral products. Huawei will jointly work with Chongqing to build a future intelligent car technology city, an intelligent super-computing center and a 5G industrial field joint laboratory. Huawei and Chongqing will launch a comprehensive and in-depth cooperation in the data center, IoT, blockchain, Industrial Internet, new smart city, smart water resources, smart terminal, 5G, smart park, machine vision, talent training and other areas.

These famous enterprises have become the barometers of Chongqing's intelligent industry and lead Chongqing's industrial innovation.

---

[1] Yan Wei, Chongqing Business News, *Alternative Fuel Vehicle Ushered in the "Dark Moment"? Car Companies Escalate Dark War in Chongqing*, July 21, 2020.

[2] Han Menglin, Xinhua News, *Ali Will Establish 20 Industrial Livestreaming Bases in Chongqing to Promote the Export Sales of Chongqing Products*, April 8, 2020.

Everything Grows with Local Innovation

Chongqing has always been an inclusive place where all kinds of whims can find the ground here to take roots.

According to the *Statistical Bulletin of Chongqing National Economic and Social Development in 2020* released by Chongqing Municipal Bureau of Statistics, Chongqing has gathered more than 7,000 big data intelligent enterprises and implemented 2,780 intelligent transformation projects, and the added value of digital economy reached 25.5% of the regional GDP.

At present, the AI industry in Chongqing is in the golden stage of accelerated development. Local enterprises have already had their own accumulation, carrying the flag of the industry of artificial intelligence of Chongqing.

In March 2020, Chongqing Chang'an Automobile and MaShang Consumer Finance were shortlisted by China's Ministry of Industry and Information Technology (MIIT) as the "Key Enterprises of New Generation Artificial Intelligence Industry Innovation" with their innovative projects in various fields.[1]

Chang'an Automobile's "Intelligent Cockpit Domain Control System" project has been built on the basis of artificial intelligence, big data, cloud computing, Industrial Internet and other multidisciplinary cross-boarding system. It includes "data platform + computing platform + 3 open containers." It could achieve customized and intelligent services so as to ultimately build an intelligent all-scene mobility ecology and meet the needs of users' personalized experience. It has provided a reference for peers in the field of smart transportation.

Mashang Consumer Finance Co., Ltd. is a technology-driven financial institution. It has more than 220 patents and developed more than 700 sets of independently-developed core systems[2]. Among them, the project "Intelligent Text and Voice Customer Service Robot Platform Based on Emotional Spectrum and Multi-Context Perception" is one of the important achievements. The project has adopted cutting-edge human-computer dialogue system architecture and is based on core algorithms including deep learning, migration learning, reinforcement learning and statistical learning. It has realized core modules such as intent recognition, intent prediction and user profiling, and built an intelligent service robot that can respond to customer questions 24 hours a day in multiple terminals and channels.

Driven by Chang'an Automobile and Mashang Consumer Finance, Chongqing local enterprises have set off a wave of innovation among themselves. A number of innovation pioneers such as Humi.com, Yaoluchuan Technology and Yunsheng Tecchology have

---

[1]Xie Li & Wang Dan,Chongqing Liangjiang New Area, *Several Enterprises in Liangjiang New Area Joined the "National Team" of the New Generation AI Industry Innovation*, March 3, 2020.

[2]Xie Li & Wang Dan,Chongqing Liangjiang New Area, *Several Enterprises in Liangjiang New Area Joined the "National Team" of the New Generation AI Industry Innovation*, March 3, 2020.

become the biggest support of Chongqing on its way of constructing a smart city and embracing the smart era.

Aiming at Locking the First Class in the Intelligent Era

Chongqing's opportunities have come along with the Smart China Expo. From the landing of famous enterprises to driving the rapid development of various industries, Chongqing only took three years.

At the 2020 Smart China Expo Online, the sparks of colliding ideas will gather strong intellectual wealth, outline a new blueprint for the development of the world of smart technology, give out the voice of Chongqing, and contribute strength to big data intelligence fields including smart manufacturing, 5G technology, Industrial Internet, artificial intelligence for Chongqing, China and even the world.

Ren Yuxin, chief operating officer of Tencent, said, "With the integration of the digital world and the physical world, production and life are being reshaped by digitalization." [1]

At present, Chongqing has initially formed a whole industry system integrating R&D, machine manufacturing, system integration, parts and components, and application services. There are many achievements in IoT, intelligent robots, cloud computing, big data, artificial intelligence and other fields.

Chongqing also focuses on the development of Industrial Internet, promotes the intelligent transformation of traditional industries, accelerates the formation of the whole industry chain of "chips, screen, smart terminals, core elements and network" and the whole factor cluster of development advantages "cloud, Internet, data, algorithm and application," and builds the whole smart scenes of "community, government services, transportation, healthcare, tourism and shopping," guiding all kinds of enterprises to the cloud to use the data, empower intelligence and enhance the new power source of economic and social development.

The development of intelligent industry in Chongqing has set up clear strategic objectives. The construction of "smart manufacturing town" and "smart city" is moving forward, aiming at the first class in the intelligent era in advance.

The main bargaining chips for the first-tier innovative status come from more than 7,000 companies in Chongqing's smart industry. The scale effects formed by these companies are attracting the attention of the world. Chongqing displays the desire and pursuit of the intelligent era.

---

[1] Sun Lei, Shangyou News & Chongqing Business News, *Ren Yuxin: Pure Offline Life and Traditional Industries Will Not Exist,* September 15, 2020.

Section IV

A City of the Future: A Global Smart City in the Era of Intelligent Industry

> Teaching is a long road, and example is a shortcut.
>
> —Lucius Annaeus Seneca

A city, more or less, deep or shallow, can be marked with multiple labels and imprints.

Chongqing is certainly the same. Hot pot, cars, beautiful girls, 8D landscape and even the subway through the building... Different people will see different angles of Chongqing city and put different labels for the city in their minds.

However, whoever revisits Chongqing will remember the "smart city" imprint that is both fresh and profound.

Chongqing has left the world a distinct impression on the road towards smart city.

Chongqing Is Committed to Participate in the Global Intelligent Process

Back in 1956, at a conference held by Dartmouth College, a group of scientists gathered to brainstorm and computer scientist John McCarthy proposed the term "artificial intelligence." Hence, artificial intelligence was officially born.

With the development of artificial intelligence, people have a new outlook and expectation for the future, global intelligence is on the horizon, and the pioneers are engaged and running towards the future.

Chongqing was previously known as an industrial city, which seemed to have nothing to do with the cutting-edge words of artificial intelligence and smart city.

The most pleasant thing in the world is always constructing an extremely high building from the ground.

Chongqing has made it, with the fastest speed, and achieved the transformation "from industrial city to smart manufacturing town and smart city." The annual Smart China Expo has become the important driving force as well as the brilliant achievement.

Although the 2020 Smart China Expo Online has been switched to online participation because of the COVID-19 epidemic, the influence remains the same. The

Chapter 5 Chongqing: A Key Witness of the Intelligence Era

Smart China Expo is leading a new round of technological revolutions and industrial changes in Chongqing and even China from production workshops to construction sites, from logistics and distribution to smart communities.

Chongqing is speeding up the implementation of the achievements of the Smart China Expo.

Nowadays, in most of the above-scale manufacturing enterprises in Chongqing, heavy and tedious work are being completed by robots and the manual assembly work is being replaced by "remote consultation" touch screen operation. Smart production lines, smart logistics vehicles and other equipment have made production more efficient and cost-saving. Chongqing's economy is entering into a high-quality development "fast lane" with the support of big data intelligence,

In 2020, Chongqing's smart industry delivered a satisfactory report card.

In 2020, the intelligent industry structure of "chips, screen, smart terminals, core elements and network" is becoming more and more perfect, and a number of key projects including integrated circuits, alternative fuel vehicles, intelligent connected vehicles, smartphones will be implemented, and the sales revenue of intelligent industry will increase by 12.8%. The construction of digital workshops and smart factories has been accelerated, 1,297 intelligent transformation projects have been implemented, and 210 municipal-level demonstration digital workshops and smart factories have been certified. As of May 24, 2020, the total number of national top-level node (Chongqing) identification registrations reached 631 million, with 334 million cumulative resolutions, access to 19 secondary nodes and 1,009 enterprise nodes.[1]

At the same time, Chongqing's 5G network, Industrial Internet, big data center and other new infrastructure projects have also accelerated, accelerating its transformation from old to new dynamics and has become a bright landscape in the global smart industry.

Responding to the Intelligent Era with Industrial Internet and Big Data

The achievements of Chongqing's intelligent industry have attracted global attention.

All inventions and creations are made for the betterment of our lives.

When people want light, there are electric lights; when people want to cross the far end of the world, there are trains and airplanes. This time is no exception, people want to have a comfortable life, so there are a variety of smart products.

When it comes to smart products, most people are more concerned about the smart attributes of the product itself, while ignoring the intelligence of the manufacturing process behind the product. In a new era of intelligent industry, intelligent

---

[1] Chongqing Economic and Information Commission, *Overview of Industrial and Information Development in Chongqing in 2020*, April 6, 2021.

manufacturing often plays a greater role in analyzing the smart products that consumers can experience.

In Chang'an Automobile's Yubei factory, hundreds of robotic arms keep waving, and all kinds of smart equipments such as automatic gluing machines, automatic equipping machines and image precision locators on the production lines are in unmanned operation. In Chongqing Jinkang Seres Liangjiang Smart Factory, more than 1,000 robots have replaced the traditional production lines, the key production lines are all intelligent, and the whole process is controlled by only a few technicians through the screen.

"Intelligent transformation" has been in full bloom in the manufacturing industry. Today, one out of every 3 laptops and every 10 mobile phones in the world is "Made in Chongqing." [1] The automobile industry that Chongqing is proud of has also changed from the once "sturdy and rough" to today's "smart and fine."

The manufacturing industry has been given a new lease of life in Chongqing, and the Industrial Internet, which is closely linked to this, has also taken a new step. At present, the development of smart factories, smart workshops and Industrial Internet in Chongqing has been striding forward.

In 2018, the first phase of Tencent's Western Cloud Computing Data Center was officially completed and launched for trial operation, which is one of Tencent's self-built large-scale data center clusters in the layout, and has become an important data center and network center for Tencent in the southwest region. At present, the second phase of Tencent's Western Cloud Computing Data Center is mostly completed. Upon its completion, it can effectively drive Tencent's Western Cloud Computing Data Center to form a computing capacity of more than 200,000 servers and become the largest single data center in western China.

In June, 2020, *Action Plan for the Construction of Major Projects of New-Type Infrastructure* in Chongqing was introduced[2]. The plan clearly states that in three years, Chongqing will invest a total of nearly 400 billion yuan to implement and reserve 375 major new infrastructure projects. Chongqing University of Posts and Telecommunications is building a big data intelligent experimental field, and has joined hands with Inspur Group to land an AI innovation platform to create a big data-oriented technology research and development, and test verification environment, in which "computing power" is the core base.

The Chongqing Big Data Achievement Transfer and Transformation Center, which

---

[1] Zheng Sanbo, Shangyou News & Chongqing Business News, *One Out of Every 3 Laptops and Every 10 Mobile Phones in the World is "Made in Chongqing"*, May 22, 2020.

[2] Zhang Hanxiang, Shangyou News & Chongqing Morning Paper, *Chongqing Plans to Invest 398.3 Billion Yuan in New Infrastructure by 2022*, June 21, 2020.

was officially licensed and established at the end of 2020, will focus on the development of the city's big data industry and the needs of big data enterprises and build a resource-sharing platform. Relying on the resources of this platform, the center will select a batch of scientific and technological achievements with high technological maturity and good market prospects, and establish a high-value achievement database. The center will also carry out offline and online combination of scientific and technological achievements release, technology supply and demand docking, and promote the transformation and implementation of the applications of domestic and foreign big data achievements in Chongqing.

The Ultimate Conception of the Smart City

In edge computing rooms, cloud computing centers, government services centers, IoT bases and smart workshops, the foundation of Chongqing's smart city has been built, and the whole city is roaring with intelligent transformation from within.

The conception of the wisdom city is quietly standing in Lijia Smart Park of Chongqing, examining every step of the future.

Lijia Smart Park, which was officially opened in August 2019, has built a never-ending smart stage.

The second phase of the park, Lakeside Lane Digital Experience Center, is once again stunning the world with more advanced black technologies, smarter park operations, richer smart life experiences and more intelligent robots, making this "tree of wisdom" even stronger.

The manhole covers, street lights, cameras, environmental testing, parking lots and other equipment and sites in the Lakeside Lane Digital Experience Center all realize IoT perception and intelligent management, fully demonstrating the interconnection of all things.

The 5G network has achieved full coverage in the park, and the 6th generation Wi-Fi network has been built in key areas, making it an outpost for the Internet of Everything.

Lijia Smart Park, with colorful culture and amazing technology, has become an Internet-famous destination.

In the "Smart Transportation" experience area, visitors can experience the feeling of automatic driving while sitting in the cockpit of a future car. In the "Smart Culture and Tourism" area, visitors can put on MR glasses and tell their destinations, and popular attractions such as the Liziba Monorail and Yangtze River Cable Car will leap to their eyes. In the "Smart Medical" area, the visitors will be able to experience the future of AI treatment room, where patients can experience the whole life cycle of smart medical solutions.

From the moment we step into the Lijia Smart Park, the future is already here. All the scenarios we envision about the future are seen in the park accordingly.

This gives confidence to global intelligence and guidance to the intelligent era. After following the map, the future is predictable. It is the process of turning a series of question marks into exclamation points, as well as the meeting of expectations and surprises.

That is also the meaning of a city of innovation to explore a great era.